ROUTLEDGE LIBRARY EDITIONS: POLLUTION, CLIMATE AND CHANGE

Volume 18

POLLUTION IN THE AIR

POLLUTION IN THE AIR
Problems, Policies and Priorities

R. S. SCORER

Routledge
Taylor & Francis Group

LONDON AND NEW YORK

First published in 1973 by Routledge & Kegan Paul Ltd

This edition first published in 2020
by Routledge
2 Park Square, Milton Park, Abingdon, Oxon OX14 4RN

and by Routledge
52 Vanderbilt Avenue, New York, NY 10017

Routledge is an imprint of the Taylor & Francis Group, an informa business

© 1973 R. S. Scorer

British Library Cataloguing in Publication Data
A catalogue record for this book is available from the British Library

ISBN: 978-0-367-34494-8 (Set)
ISBN: 978-0-429-34741-2 (Set) (ebk)
ISBN: 978-0-367-36518-9 (Volume 18) (hbk)
ISBN: 978-0-367-36527-1 (Volume 18) (pbk)
ISBN: 978-0-429-34670-5 (Volume 18) (ebk)

Publisher's Note
The publisher has gone to great lengths to ensure the quality of this reprint but points out that some imperfections in the original copies may be apparent.

Disclaimer
The publisher has made every effort to trace copyright holders and would welcome correspondence from those they have been unable to trace.

Pollution in the Air

Problems, Policies and Priorities

R. S. Scorer

Professor of Theoretical Mechanics
Imperial College, London

Routledge & Kegan Paul
London and Boston

First published in 1973
by Routledge & Kegan Paul Ltd
Broadway House, 68–74 Carter Lane
London EC4V 5EL and
9 Park Street,
Boston, Mass. 02108, U.S.A.
Printed in Great Britain
by Ebenezer Baylis & Son Ltd
The Trinity Press, Worcester, and London
© R. S. Scorer 1973

ISBN 0 7100 7569 3

Contents

Preface

All around us pollution is on the increase: in the air, in the water, on the land. It is mainly due to the use of industrial technologies and the enormous turnover of material in our cities. Pollution has grown as the population has grown and particularly where the people have become prosperous.

Very soon we shall approach close to the pollution barrier – the limit beyond which there will be irreversible catastrophe. At the same time we ever more rapidly grow towards the population limit – beyond which further increase necessarily means a lowering of all standards of life. In both respects this will be a new experience as we find the world around us too full of one another with no room for manœuvre. New attitudes and values will be needed if civilisation is not to degenerate into squalid anarchy.

To live we need our technology, and that means pollution! Or does it? Is there a way out? Can we solve our problems without trying to stop industrial development? Do we have to stop the growth of our civilisation to prevent that growth from becoming malignant?

These are terrifying questions and many are the prophets of doom. I believe that pessimists have a valuable role in awakening society to the need for new policies and even a new goal for civilisation. They challenge the optimists to solve the problems, and it is the duty of those who feel that hope is justified to participate in finding our way through the severest test that evolving man has ever faced.

Nothing is certain but I think we are entitled to search for answers in further advances in technology. So far technologists have not been asked to solve pollution problems; quite the reverse, they have been asked to develop industries which make it and have been well paid for doing so. The situation is of man's making, and it is in the belief that the problem is to redirect man's effort that this book is offered.

It is quite evident that a new generation does not accept the values of its parents, and just as dangers in war have produced great social revolutions so I believe that an appreciation of the present need for an approach to pollution problems uninhibited by old values will foster new attitudes in society and a will sufficient to alter the course of civilisation.

Hitherto we have acted in response to disasters of greater or lesser magnitude. We have merely thought of stopping the grosser abuses of our environment. Now we need to give most of our attention to the problems being created by the directions in which domestic and international commerce are driving technological developments. Therefore, after setting the background, I have discussed the problems of controlling what happens. Nothing will be effective without public support, and that will only come through public discussion, and that is what this book is intended to stimulate.

I am grateful to Cape/Penguin 1971 for permission to quote from *Soledad Brother* by George Jackson, copyright © 1970 by World Entertainers.

1 The Need for Air Pollution

'The atmosphere is the bad breath of life.'

An atmosphere of pollution

Every breath pollutes the air. Every corpse or decaying leaf emits poisonous odours. Now that man, animals, and vegetation have been joined by cars, ships, aeroplanes, houses and a host of industries as polluters of the air we are worried. Yet if we look at our world we find that in one sense our whole atmosphere is pollution. We need to accept this viewpoint because although our way of life has introduced polluters on a quite new scale in this century, the atmosphere is not one of absolute purity. The one we are fouling is an atmosphere whose very composition has been determined over geological time by the unwanted emissions of life and death on earth.

The lifeless planets

The Earth's atmosphere is unlike that of any other planet. The Moon and Mercury have no atmosphere because their gravitational pull is not enough to hold gases to their surfaces. Mars has a very tenuous layer of gases on its surface, with very little water and other compounds of hydrogen, and its winter polar 'ice' cap is probably made of carbon dioxide crystals ('dry ice') such as we use in cold boxes.

Venus has succeeded in retaining about a hundred times as much atmosphere as the Earth even though it is smaller, and the result is that it is very hot on the solid surface of the planet. It is too hot for any life to be there. It is also very dark because the whole planet is completely covered with clouds. We know from this fact that the clouds are not made of water like clouds on Earth, but are composed of ammonia or methane or some other substance which does not condense easily into droplets of rain like water does.

Jupiter and Saturn are so large that they have much denser atmospheres than Venus. Jupiter in particular cannot possibly support any life because the temperature and pressure at its solid surface are much too high, and it must be very dark there indeed. The outer planets – Uranus, Neptune, and Pluto – are very much further away from the sun, and are probably far too cold to be able to support life unless, like Jupiter seems to have, they too have great internal sources of heat.

The Earth's atmosphere was not created as it now is. There are carbonates in the rocks which contain a much greater total amount of oxygen than the atmosphere. All the chalk and a large part of the limestone and coral were created by living organisms which must have extracted the oxygen from the atmosphere; thus there must have been much more gaseous oxygen or carbon dioxide at some stage in geological history.

Nitrogen now composes four-fifths of the total mass of the atmosphere, yet there are thunderstorms and other mechanisms which cause it to combine with oxygen and ultimately become nitrates in the ground. It is life that has filled the air with nitrogen, and our atmosphere is quite different from those of other planets in this respect. When life began we do not know how much of the world's nitrogen was gas, but it was almost certainly very much less than today. When we learn about nitrogen in chemistry lessons we find that it is a rather inactive gas, yet all the nitrogen in the atmosphere must have been part of living substance at some time in the past, and cycling continues as fast nowadays as ever.

Carbon dioxide is only about one-two-hundredth of the atmosphere, and water vapour about one-thousandth of its mass, yet both these substances are passing all the time in enormous quantities through animal and vegetable forms of life. One of the reasons for believing that life does not exist on the other planets is that their atmospheres are composed to a much greater extent of the compounds of carbon and hydrogen such as methane and ammonia. Perhaps Venus and Jupiter are destined many millions of millions of years hence to grow life forms which will transform their atmospheres into living media with ever-changing beauty in their skies, at a time when our own planet is in its final evolutionary decline, but in terms of human life as far ahead as we can imagine they are dead planets.

The ocean contains several million times as much water as the atmosphere, and dissolved in it is much more carbon dioxide than exists in the atmosphere; there is more life in the ocean and more energy from the sun used in the ocean to create protein than on the land. The ocean is the main reservoir of biological activity which has made our atmosphere what it now is. No other planet has an ocean, or even a body of water that could be called a sea, or lake, or even a pond. At best Mars may have slight occasional dew, while Jupiter and Venus are too hot to have liquid water on their surfaces. Probably Jupiter's two largest satellites Ganymede and Callisto are almost hopeful as homes for life and I will leave them to the optimistic imagination of science-fiction writers. They will be in the news before long when space probes are improved but I fear they will be as barren as Mars and Mercury.

The living atmosphere

The Earth is a living treasure in an almost barren universe. That we humans are the latest product of its evolution does not entitle us to think of it as ours, to be done with as we will. We are but one element in the continual birth and death of species. For us to be able to live as we do in an atmosphere which offers such great beauty and opportunities there needs to continue the same old exchange of gases as has gone on for thousands of millions of years. If we upset the continual cycling of the atmosphere which is not only the source of the needs of all life forms but also the main depository of their pollution we shall find our own lives in danger. This could happen if we become too numerous or if we add to the biological cycles such a large tonnage of industrial effluents, the by-products of our affluence, that they begin to change the composition of the atmosphere and ocean.

Some pollutants which we produce may turn out to be good for our world, others may be terribly destructive. The possibilities for changing the situation for the better arise mainly from cultivation, using the cycles as they now exist. But at the same time our industry and cities are feeding ever-greater amounts of new and often obnoxious substances into the air and water and it is these we need to watch and control.

Because it is continually being renewed by life forms and cleansed by rain, the atmosphere has a great capacity for recovery from almost any damage we do to it. It is this very fact which has made the evolution of life forms so productive and viable on Earth. Nature is continually experimenting with thunderstorms and volcanoes, earthquakes, landslides, and so on, and nothing we have done artificially so far compares in magnitude with what has happened naturally. Our environment is not fragile, it is very resilient and powerful indeed. We need have no fear that the weather will suddenly turn sour on us, for long before that could happen we would have killed the life in our seas upon which we depend for our food, and the malodorous coasts would long since have forced us to change our ways. The atmosphere will be the last natural force we might damage because it is so full of motion.

Where then are the dangers?

What are smogs?

The air circulates so rapidly and is cleansed so efficiently most of the time that we get used to putting smoke and various gases into it and they simply 'go away'. But sometimes part of the atmosphere becomes stagnant for a few days, and the pollution there begins to build up into a pall, and this we call a smog. We don't like smog because we are used to pleasant skies, and yet most of us survive them without scars. Only a few very old or young or infirm are killed, and these are few compared with the daily slaughter on the roads or the terrible death rate from smoking diseases. Even suicide is a cause of more deaths than air pollution. Smogs are not a catastrophe for us comparable with war. Smogs will not seriously hurt humanity directly.

Nevertheless smogs and other kinds of air pollution cause much squalor and unpleasantness in cities. They also do very serious damage to vegetation. Insects and birds then begin to be harmed, and all biological activity in the immediate environment of the cities producing the pollution is lowered, and begins to move away. California, which used to be a land where spinach, beans and pine trees grew in great abundance, is now having progressively to give up cultivating them. Soon the peaches and figs will suffer, too, and this wealthy community will find itself living in a barren land instead of in a green paradise, importing its food from territory not yet spoiled.

Cities have always done this. The ground is paved over and built upon; industries take the place of grassland, horses and birds gradually move away except for the parasitic pigeons and starlings who like a city's winter warmth, and have learned to join with man in ruthless greed. But a city is a deliberate construction, and we hope it will remain surrounded by farms where the natural cycles are exploited, and indeed encouraged, by careful husbandry. A state or country which turns itself into a city in this sense must depend on another part of the world for its sustenance, and the world, being finite in size, can only support a certain number of such states. We have not yet reached that number, but the growth of industry and population is making us approach it at an ever-increasing rate. The trouble arises because we have a commercial and economic way of life which is dependent on growth, and unless we take steps to slow down growth we shall find that when the world is full we have no political and social organisation capable of governing our people without growth. Anarchy will then be the first catastrophe that will befall our race.

The expansion of water pollution

The second danger lies in the fact that rivers, lakes, and seas do not circulate as rapidly as the air. Consequently, where too much pollution is emitted it overwhelms the biological mechanisms which digest it normally, and the forms of life in the water are changed, and changed for the worse.

When life forms are destroyed, their environment is quickly changed because they are not there to keep it in its original form, and the result is that it is very difficult for them to return. The polluted areas therefore tend to expand and the pollution is not digested in the areas where the old life has been destroyed. Around Japan and in the Baltic Sea, the polluted areas are rapidly joining up and the fisheries progressively being ruined.

Since the present trends make it certain that humanity will make more pollution every year, the prospect is that a very much larger area of sea will be rendered useless before anything we may try to do will have any effect at all. To reduce pollution of the sea is therefore a matter of desperate urgency.

Since it is the high concentrations of water pollution that are harmful we may have to use the atmosphere as the main vehicle for carrying away unwanted material. The products will then be distributed over a very large area in small concentrations which can be digested by nature. This we already do when we incinerate our rubbish. We turn it into gases and fine dust particles and the atmosphere 'carries it away'. Because our burial grounds have coffins and products of the stonemason's 'art' to digest as well as human bodies, they are extremely inefficient at their job, and for this reason we are having to use cremation. Some cities find that the easiest way to dispose of the branches cut from the trees in their streets and parks is to burn them, and this has to be done in an expensive incinerator, not in bonfires, in order to avoid obnoxious pollution by smoke.

The destruction of species

Insecticides are a different problem altogether. They have become a dangerous form of pollution because although they are very well dispersed by the wind and by being washed down rivers into the sea, certain forms are removed from the water by plankton and other low forms of life and not deposited in the mud of the sea bed. The plankton are eaten by fishes which in turn are eaten by other fishes which in turn are eaten by yet other fishes or by sea birds or by man. These last forms of animal life are therefore described as being at the end of the food chain or at the top of the ecological pyramid, and it is in them that some of these poisonous substances become concentrated.

It is possible that some animal forms may be made extinct as a result of this because they succumb to the poison when they have consumed enough of it. There is no indication that man will suffer this fate because of any poison already loosed into the ocean but this is partly because the food we eat is subject to continual chemical analysis and we are warned in time not to eat the animal that precedes us in the food chain as soon as it becomes dangerous: we eat something else instead.

Nevertheless, it is very bad to allow any form of food to become inedible. In the Baltic Sea, mercury compounds have been found in dangerously high concentrations in some fish, and the sale of them for food has had to be prohibited. The mercury originated as a fungicide used in the logging industry to protect timber from deterioration before it reached the sawmill.

But life forms are to be treasured not merely as food. Their genetic heritage is the accumulated wealth of thousands upon thousands of years of evolution and it can never be re-created by man. By comparison with the greatest works of art they are priceless treasures, and should not be allowed to die because of man's carelessness. Such a death would be perhaps more tragic than the extinction of the tiger or the great auk which, though noble in themselves, are not the particular friends of man that the fishes are which supply one of our main sources of food.

Sulphur : food or poison?

How our attitude towards a form of air pollution can change is well illustrated by the case of sulphur dioxide. A whiff of it in the chemistry laboratory will set anyone coughing, and it can kill although it scarcely ever does. Concentrations which animals can tolerate can do serious damage to some forms of plant life. On the other hand, it is widely used as a food preservative because it prevents the growth of undesirable bacteria without being in the least poisonous. It was once commonly used as a disinfectant after a room had been occupied by a person with an infectious disease; the room was sealed, a sulphur candle burned in it, and the sulphur dioxide so produced was left for a day or two to do its work.

Sulphur dioxide became known as a pollutant because sulphur is contained in almost all coal and oil. The pungent gas put out by a coke brazier is sulphur dioxide, and we keep our distance but do not worry about breathing because, even if it makes us cough, we soon recover. It turned out to be fairly easy to measure chemically, and so there are now instruments established all over the industrialised world, recording the amount of sulphur dioxide in the air. It became a measure of the degree of pollution.

We do not use carbon dioxide or water vapour as measures of pollution because there is so much of both of them in the atmosphere already that we could not attribute them to the pollution source with any certainty, even though they are the two most plentiful components of what comes out of any chimney (except of course for the nitrogen, all of which went into the furnace from the air in the first place).

In any normal school or college textbooks about the air, sulphur dioxide is not listed as a normal component of it. It is regarded as a pollutant. But it is essential to all forms of vegetation, unless they are fed with artificial fertiliser, for there to be some sulphur dioxide in the air. All forms of life need sulphur. Sulphur dioxide is so quickly removed from the air that it is not regarded as a genuine component: its residence time is very short compared with other gaseous components. Provided that we do not anywhere produce concentrations of it that it might do harm, we can think of sulphur dioxide in the same way as carbon dioxide and water vapour. All we have done is to feed more of each of these substances into the natural cycle which already had a very large turnover. No harm is done until we increase the turnover so much that we produce a significant change in the amount of it in the atmosphere. It is again a question of the total tonnage, and as we shall see in later chapters, we have not approached dangerous levels yet.

Our need of air pollution

We thus have a picture of an atmosphere composed of substances which are emitted by some life forms and absorbed by others.

The relative amounts of nitrogen, oxygen, carbon dioxide, water vapour, sulphur dioxide, and a host of other minor substances which enter and leave the atmosphere all the time, are determined by the input and extraction rates rather than by any inherent feature of the atmosphere created with the earth itself. The whole atmosphere is somebody's pollution, and in this light we must think of any new substance we emit into it.

With knowledge we need have no fear, provided we have the political means to control man's own activities when they become excessive. We have to remember that every new activity requiring the expenditure of energy requires us to emit more pollution. It is so much safer to put pollution into the atmosphere than to let it accumulate in the water or on the land that we need to use the atmosphere's dispersive power in almost all our enterprises.

Lest anyone should think that what has been said provides a justification of air pollution it must be emphasised that all pollution which we need to emit ought to be harmless in every respect. We should not be able to see it or smell it; we should not experience any damage from it; the air should always be as fresh and clear as it was before the Industrial Revolution; technology is such that it need not be otherwise if we wish it. Obviously today much of the atmosphere is unpleasantly and obnoxiously polluted, and it is a great challenge and a gigantic task to stop this, but it certainly does not mean that all pollution must be stopped to achieve this end.

2

2 Cleansing the Air

The residence time

We have already seen in chapter 1 that sulphur dioxide has been regarded as a form of pollution because it is so rapidly removed from the atmosphere that large concentrations of it are rare. Carbon dioxide, on the other hand, is produced by animals and by the world's industries in far greater amounts than sulphur dioxide but is not bothered about because what is added is only a small fraction of what is there already, and so there does not seem to be any difference. We are quite capable of living in an atmosphere with up to ten times as much carbon dioxide in it as our atmosphere, and many people do so from time to time indoors when a room is crowded or badly ventilated; but the mixing up of all gases released out of doors is so rapid that such large concentrations seldom occur there.

During the infamous London smog of 1952 when the air became very stagnant over London for more than four days, it was estimated that the air contained about ten times the usual amount of carbon dioxide. People's breathing and the burning of fuel used up about one-sixth of the oxygen, but no one noticed any difference: we got used to the carbon dioxide and the decrease in oxygen in each breath was equivalent to going up a mountain about 1,700 metres high. The effects of smoke and low temperature were much worse than any other effects.

We can calculate how rapidly the oxygen was replaced by carbon dioxide because the residence time of both is large. If no carbon dioxide were created it would take a few years for it all to be used up. If no oxygen were fed into the atmosphere, again it would be a few centuries before it was used up, assuming that the animals and all the world's fires continued to consume it at the present rate. On average, a molecule of oxygen remains in the atmosphere without change for several years, and a molecule of carbon dioxide for quite a few. An average molecule of nitrogen probably stays in the air for several centuries before it is combined into some organic or inorganic chemical compound to be returned to the atmosphere ultimately after a spell of duty in the biological world. The residence time of sulphur dioxide on the other hand is a few hours, and so most high concentrations of it are found fairly close to where it was emitted into the air, again causing us to think of it as pollution.

But we must not think in a simple-minded way that every kind of chemical has a residence time which we could display in a table. Such a table would look something like this:

argon	for ever
nitrogen	10–1 million years
oxygen	1–1,000 years
carbon dioxide	1 month–100 years
sulphur dioxide	1 minute–1 month
fluorine compounds	1 minute–1 week
ammonia	1 minute–1 week
smoke	1 hour–2 months
oxides of nitrogen, hydrocarbons, and other odours; dust, sea salt etc.	1 minute–2 months

There is no precision at all. Even if we knew precisely what the average residence time for a substance was, it would be of no help because we can think of circumstances in which molecules could remain for an interval anywhere within the range shown in the table.

If sulphur dioxide is emitted into a closed room, in an hour or two nearly all of it will have been absorbed onto the walls and other objects such as clothes, furnishings, and books, and in a smog situation almost all the objectionable substances are so rapidly absorbed indoors that even though it becomes stuffy indoors it is still much more pleasant than outside.

Outside, in calm air, pollution is absorbed onto buildings, trees and so on, but more slowly than indoors because there is less solid surface close to a given volume of air. Pollution emitted by an aircraft remains in the air longer than pollution from a car simply because it is further away from the ground and therefore takes longer to reach it. We must not argue that aircraft pollution is therefore worse because it remains in the air longer. It depends entirely upon who or what is damaged by it. Car exhaust is often emitted in enclosed streets where it is breathed, or comes into contact with flowers, soon after being emitted. Car exhaust in a country lane is a negligible problem because there is so little of it and so few people to be bothered by it. Aircraft exhaust at a height of 10,000 metres is no problem at all because by the time anything, even a bird, makes contact with it, it is so mixed up and diluted with the surrounding air that it cannot even be detected.

Aircraft flying in the stratosphere well above 10,000 metres raise some new problems because the residence time is much longer there. These are discussed in more detail in chapter 5.

There are many other substances such as the smell of pinewoods and desert dust which are major pollutants in the atmosphere in the sense that their tonnage is comparable with that of many other pollutants which we worry about. But we shall not discuss them in detail. All chemical compounds with large molecules, such as vapours which we can smell, and all solid substances, such as particles of sea salt and dust blown up from the ground, are removed very efficiently by a variety of mechanisms which we are about to discuss. What is important is that all the obnoxious gases have short residence times. It is their readiness to engage in chemical reactions which makes them obnoxious, and also ensures their rapid removal. Almost all solid particles which are smaller than 10 microns (10 thousandths of a millimetre) in diameter act as condensation nuclei or are fairly quickly caught up in droplets of water when cloud is formed. Cloud also absorbs gases with large molecules or which are chemically active and so if a particle is not absorbed at ground level it has to wait until it is brought down by rain, snow, hail, or drizzle. The most important factor in the residence time, therefore, is how soon does a particle, or molecule of gas, find itself inside a cloud.

Since it scarcely ever rains in the Sahara desert, fine dust that is blown up into the air there often has to travel far beyond the boundaries of Africa before it comes again to earth. Occasionally it comes to earth in rain on Britain or the United States of America, but most of it comes down much nearer to Africa. Nevertheless, because it has to travel quite a long way before it comes to a rainy area it often remains in the air for five or six weeks.

By contrast, a particle of salt blown upwards from the sea surface in the equatorial rain belt area could return to the sea in a rain shower within as little as twenty minutes. Smoke emitted from a chimney into a drizzly fog could equally quickly descend to earth. Sulphur dioxide put out by an oil refinery in a desert country of the Persian Gulf could travel across arid lands for as long as Sahara dust before being removed from the air, whereas about half of the sulphur dioxide that is emitted from the chimneys of Britain descends on to Britain itself or very close by, and this means it reaches the earth within a day or less.

Where is most of the pollution removed?

This question is not asking whether the pollution is removed from the air over London or over the Atlantic Ocean, but whether most of it is captured in the lowest layers of the atmosphere. Or is there a large reservoir of pollution in the stratosphere above a height of 10 miles from which it is gradually extracted?

To obtain an answer to this question, it is no use saying that it is terribly complicated and almost anything can happen even though that is more or less true. Equally, an average residence time is of little use as an answer. We have to make a compromise between oversimplification and bewildering complexity. It is worth noting that many scientists have found themselves collating more information than they can make use of nowadays. Automatic instruments are set working and data is continuously accumulated. In order to digest this computers are set working, and at this point we need to remain very clear headed. If we tell the computer what sort of analysis to make of the voluminous data, some suppositions are needed about how the data ought to be sorted: we need to know something about the answers before we have any. In many cases we find, if we find anything, that it would have been better to design the instruments and find sites for them in a different way from the one we first chose, and the moral of this is that we have to advance at a rate at which the human brain can comprehend each successive new step as it is made. We cannot make a great leap forward into the unknown by computer. Computers have no knowledge or intelligence: they are simply quick-working sorting machines and calculators and they cannot be endowed with any of the attributes of imagination. We must not imagine that any of the problems of pollution can be solved by computer. There is no short cut to clear thinking, and no substitute for it.

The four layers of the atmosphere

In dividing the atmosphere up into four layers which are relevant to pollution, therefore, we do it to help us in our thinking and decision-making, not because these four layers are all the time distinct and separate. They are:

1. The air close to the ground from which pollution can be absorbed directly on to the vegetation, buildings, water surfaces, and so on. Depth 1 metre to 100 metres.

2. The layer up to the base of the clouds at somewhere between 500 and 2,000 metres.

3. The cloud layer, from the cloud base up to the tropopause, which is at around 10 km above the ground, but could be as high as 17 km near the equator or as low as 6 km nearer the poles.

4. The stratosphere, above the tropopause, where there are scarcely ever any clouds.

The ground layer reminds us that substances like sulphur dioxide are absorbed from the air as it flows over a water surface or over grass, trees, and buildings. Trees are very good absorbers of sulphur dioxide, and a good hawthorn hedge in the growing season can remove well over half the sulphur dioxide from the air blowing through it. Most of the air in the wind goes over the top of the hedge, of course, and so a hundred yards or so beyond the hedge the ground air has again become so mixed up with air higher up that the concentration of sulphur dioxide on the upwind side of the hedge is almost restored.

Nevertheless, probably one-third of the sulphur dioxide removed from the air is cleaned out of it directly from this lowest layer.

The layer up to cloud base is a turbulent one. Pollution is nearly all produced at the ground, and the effect of this layer is to stir it upwards until the pollution is more or less uniformly mixed up to the level of the cloud base. Almost no pollution is removed from this layer directly except by rain or snow falling through it, but even that does not remove more than a small fraction. Drizzle and wet fogs are much more efficient at cleaning the air. A fog close to the ground is part of the ground layer, while drizzle usually falls from cloud higher up.

Clouds are formed in air which rises from below through the level of the cloud base. The more rapid the ascent of the air the more likely it is that almost every solid pollution particle will become the nucleus of a cloud particle. All the gases which are soluble in water, such as sulphur dioxide, quickly begin to be dissolved in the cloud droplets. If the pollution is chemically active quick reactions take place in the droplets. Thus, sulphur dioxide becomes sulphuric acid, or ammonium sulphate if there is any ammonia about.

Ammonia, like nitric oxide, is formed naturally and is continually fed into the air. Because it is as active chemically as sulphur dioxide it is at least as quickly removed.

Many clouds which are formed evaporate again within ten minutes to one hour, and the pollution is released into the air into which the cloud evaporates. Thus some of it is mixed into the air above cloud base; but almost none is carried from below to above the cloud base level except in an upcurrent forming a cloud. Consequently, the air above cloud base is much cleaner than below, and it is very easy to see from an aeroplane that this is so.

Many clouds produce rain, hail, snow, or drizzle, and this falls to the ground carrying pollution with it. The air that contains the cloud from which water, laden with pollution, is falling will ultimately begin to descend again. Because it has lost water, the cloud that remains in it will evaporate long before the air gets down to the base of the cloud again and it then becomes an invisible part of the layer above cloud base. Since it has been cleansed, most of the air in that layer is very noticeably cleaner than the air below cloud base.

The motion of the clouds is very vigorous, and so all the air up to the highest level reached by clouds is well mixed, and frequently cleansed of pollution. All the air up to the tops of the highest clouds (except for a few clouds which temporarily protrude into the layer above but soon fall back again) is well mixed, and is called the troposphere (mixed sphere). Its top is called the tropopause, because at that level the mixing almost stops and above lies the stratosphere (layered sphere) which is not at all well mixed. There is very little pollution there because although there is some exchange across the tropopause, most of the air going upwards to become part of the strato-sphere has already been well cleansed by producing rain or snow.

It is very rare for air in the troposphere to go without being cleansed for more than a month, and so the only substances that remain for longer are those that are not mostly washed out by rain. Carbon dioxide is an example of this.

The stratosphere is composed mainly of a mixture in the same proportions of all the components of the air that exist in the troposphere except those that are washed out by rain. Any differences that exist are due to chemical changes in the stratosphere or the direct introduction of pollution from below such as by a volcanic eruption. Of all the chemical changes the most important is the production of ozone, which takes place at heights between 30 and 60 km above the ground. Ultra-violet light decomposes molecular oxygen (O_2) into atomic oxygen (O). Some of the oxygen atoms combine with molecules already containing two atoms to form ozone (O_3). The ozone is also decomposed by the ultra-violet light but absorbs it in the process, and therefore prevents it from reaching lower levels in the atmosphere. The layer 30–60 km above the ground is continually forming and destroying ozone. Some of it gets down to lower levels where it is protected from ultra-violet light by the ozone above, and so it lasts much longer. In fact most of the ozone, for this reason, is at much lower levels in the air and only reverts to the more common diatomic oxygen (O_2) after a month or so in the lower strato-sphere. It may last for a year or so lower down in the troposphere if it does not participate in an ordinary chemical reaction. In the ground layer, ozone disappears very quickly by chemical reactions on surface objects and with short-lived forms of pollution found near the ground.

Any pollution put into the stratosphere by a volcanic eruption or nuclear explosion may remain there for a few years, the time being longer the higher up it is originally introduced. It is only removed by an exchange of air across the tropopause, and then it is, of course, very soon washed out after the air enters the troposphere. Roughly we can say that in the bottom 2 km of the stratosphere pollution could remain for a few months, but for a time gradually increasing to about five or ten years at a height up to about 30 km. Above 30 km there is only about one-thirtieth of the total mass of the atmosphere, and so it is not a very important reservoir of pollution except for the ozone which is produced there.

Chemical transformations

When pollution is emitted into the atmosphere it may remain unchanged until it is finally removed. Dust is like this. But it is not so with almost all gases. Chemical reactions involve some of the nitrogen and oxygen, as well as the more obviously reactive substances like ammonia.

Some of these reactions are of a rather straightforward kind such as could be demonstrated in a school chemistry laboratory, while others are of a much more special kind and take place at low pressures and concentrations, in the presence of water, on the surfaces of solid particles, under the action of sunlight, and so on.

The following are fairly straightforward, although it is not obvious how they actually take place in detail:

hydrogen sulphide (H_2S) + oxygen (O_2) becomes water (H_2O) + sulphur dioxide (SO_2)

SO_2 + oxygen becomes sulphur trioxide (SO_3)

or, SO_2 + oxygen and water becomes sulphuric acid (H_2SO_4)

SO_2 + oxygen + ammonia (NH_3) in the presence of water become ammonium sulphate (($NH_4)_2SO_4$)

carbon monoxide (CO) + oxygen becomes carbon dioxide (CO_2)

oxygen and nitrogen (N_2) become oxides of nitrogen (denoted
   by NO_x because there are several: N_2O, NO, NO_2 etc.)
hydrocarbons (C_xH_y) + oxygen become carbon dioxide and
   water.

The characteristic of these reactions is that stable compounds
are formed from chemically active substances. Hydrogen
sulphide is the unpleasantly smelly gas emitted by rotting
vegetation and animal residue. It is one form in which sulphur
mainly enters the atmosphere from the animal and vegetable
world. But it soon becomes sulphur dioxide, which in turn soon
becomes some form of sulphate. Sulphur trioxide takes the
form of very small solid particles which, when plentiful, have
the appearance of a bluish smoke. It is not easy to capture
and does not very readily form sulphuric acid by combination
with water vapour, but remains as very tiny solid particles.
It seems likely that it is formed mainly inside flames, or at
high temperatures in furnaces, and not in nature.

Sulphur dioxide is, however, quickly transformed into
sulphuric acid inside water droplets, and this happens very
readily in nature. Some chimneys emit quite a lot of SO_3
which is presumably formed at high temperatures; others emit
droplets of H_2SO_4 which is presumably produced at much
lower temperatures in flue gases, and sometimes in the atmosphere after they emerge from the chimney.

Ammonia is emitted in very large quantities from animal
urine and excreta, and quickly combines with SO_2 and oxygen
in the presence of water (probably more slowly when only
water vapour is present), to form particles of ammonium
sulphate, which is an excellent fertiliser. Thus if it falls on the
ground and is fed to the roots of vegetation by rain it will
promote growth. Ammonium sulphate particles are found in the
air over land all over the world.

Carbon monoxide is emitted by almost any furnace or internal
combustion engine in which the combustion is not complete.
But it is soon oxidised to harmless carbon dioxide and so it
does not accumulate in the air. It should be noted that although
it has the property that it prevents the blood from taking up
oxygen, and can cause death when present in large quantities
as in a closed garage, it is no danger in streets.

Oxygen and nitrogen are the two main components of the

atmosphere and they combine to a small extent wherever the temperature and pressure are high enough. They combine in flames, inside car engines, in lightning, in volcanic eruptions, in the trails of meteors (shooting stars) and also in forest fires, and even in central heating plants and cigarettes to some extent. There is therefore a small quantity of oxidised nitrogen about in the atmosphere all the time. It ultimately becomes transformed into nitrates, which are good fertilisers, and is rained out of the atmosphere.

The special interest in NO_x is that its greatest concentrations are produced in cities by car engines, about which we shall have more to say later. For the present we merely note that because these compounds are fairly active chemically they do not accumulate in the atmosphere, and are only found in harmful concentrations in stagnant air containing a large car population.

Hydrocarbons in the atmosphere originate mainly from oil fuel, gasoline, and other solvents, by evaporation. From cars about 4 per cent of the fuel comes out of the exhaust unburnt, and this is the other main artificial source. Nature produces enormous quantities of these substances, perhaps the most obvious being the smell of pinewoods.

Exactly what happens to hydrocarbons depends upon the circumstances. There are probably hundreds of different chemical reactions in which they participate, the end result of which is to turn them into carbon dioxide and water vapour or into a variety of solid or liquid substances which produce a haze in the atmosphere. These are produced through the intervention of oxides of nitrogen and sunlight. The blue haze seen over a forest in sunshine is as much a photochemical smog as the smog of Los Angeles, but it is much less harmful because it is less concentrated. Indeed when the materials have been transformed into a haze all their damaging properties have been eliminated and it is mainly because of the continuous production of large quantities of the raw materials from which the photochemical haze is produced that it is so unpleasant. The smarting of the eyes and the damage to materials and vegetation caused by a photochemical smog are due to substances present while the transformation to the haze is taking place, and our problem is solely the high concentrations reached in an automobile-based conurbation.

The picture which emerges from all this is one of great chemical activity within the atmosphere itself. In addition there is the continuous exchange between living things and the atmosphere, and the absorption by them of substances removed from the atmosphere by rain.

The whole system is quite easily capable of absorbing much more output of pollution than is emitted by the activities of man at the present time. Where then does the danger lie?

The answer to this question is quite simple: pollution only becomes dangerous when there is too much of it in one place. In the past some industrial processes have been managed so inefficiently that a long-term form of pollution has been produced. The best known of these occurred in the non-ferrous metal industry of Swansea, where in addition to the high concentrations of SO_2 and SO_3 that were produced from the large amounts of sulphur in the ores used, large quantities of chemical compounds of the metals themselves were deposited on the ground near to the works. The result is that although the vegetation was initially destroyed by the sulphur oxides, its rehabilitation has been prevented to a large extent by the poisoning of the ground by the metals. This 'poison' is not washed out of the ground by rain and can only be removed satisfactorily by growing selected grasses to which the metals are not toxic for twenty years or so. During this time, these areas cannot be used for grazing by animals because of the toxicity of the metal compounds. The only alternative remedy is the replacement of the top soil.

This is an example of the carelessness of industrial activity in the Victorian era. Many streams, and occasionally some urban water supplies, have been poisoned by crudely managed smelting activities, particularly in parts of Wales and Scotland.

Smogs

This word has an almost humorous ring about it, and as a consequence it has been used as the almost friendly name given to forms of air pollution by people who live in smogs of their own making.

London smogs culminating in a very unpleasant incident in December 1952 had been famous for decades. Although it has often been alleged that the Great Smog of 5–8 December 1952 killed 4,000 people it is important for other reasons. London had suffered from 'pea soup' fogs consisting of a pall which often made the city very dark during many winters, and an attitude of resignation to them as natural phenomena had grown up. The National Smoke Abatement Society had analysed their cause and proposed a remedy at the turn of the century but not enough people regarded the matter as urgent enough for public action. It is true that 4,000 deaths could be attributed to the Great Smog of 1952 in the sense that it was their immediate cause, but almost all of the adults who died were in a very weak state of health and the next influenza epidemic or period of weather leading to severe colds and occasional pneumonia would have claimed them as victims if the Great Smog had never occurred. Probably half of them would have died as a result of Christmas activities if they had lived another three weeks, for that is a normal 'peak period' in the deaths of the infirm aged.

Before the Great Smog there had been a period rather freer than usual from deaths from respiratory diseases although there had been incidents which had weakened many people with chronic bronchitis. It was a combination of circumstances that led this particular smog to apply the *coup de grâce* which these infirm people were waiting for; the run of events merely concentrated their deaths into that particular week instead of allowing them to be spread more evenly through the winter. A minority of the victims were newly born babies who, along with some prize farm animals assembled at the Smithfield Show, died because they had no adequate protection mechanisms to meet this first smog of their lives.

This Great Smog, however, was also the last straw which made the public ready for action by the government to prevent the recurrence of such disastrous incidents. Even then the government cannot be said to have taken a very firm initiative because it merely arrived at last at the view that it would not become unpopular if it allowed action to be taken. Undoubtedly, too, the continual repetition of the misleading statement that the Great Smog of 1952 killed 4,000 people (as if they had been a sort of random cross-section of the population suddenly struck down in the street by this terrible event) helped the campaign for clean air enormously. I rode many tens of miles through it on my bicycle: my eyebrows, nostrils, and clothes generally became quite filthy. Curtains and other furnishings got very dirty and a haze could be seen in the air indoors; but no ill effects were experienced by normal healthy people, although it was a most unpleasant experience.

London's smog was, as the name implies, a mixture of smoke and fog. The smoke was produced by the inefficient burning of coal, and the fogs were a natural phenomenon. Smogs no longer occur because the smoke is not now emitted. Furthermore, although London is a place where natural fogs are likely to occur very frequently, the paving over of streets to an ever-greater extent has caused the rain to be drained away instead of being absorbed in the ground as in a sponge. The consequence is that there is much less evaporation from the ground and the humidity is lower than it used to be. Natural fogs are less common, and are characteristically concentrated in the large parks and open spaces of London, where the vegetation emits large quantities of water vapour into the atmosphere and keeps the temperature down below that of the built-over areas.

It is important to appreciate what smoke really is. Open fires produce it because coal and wood emit combustible vapours when they are heated. These vapours are often not ignited and go up the chimney condensing into solid particles of smoke as they go. They, together with fine particles of ash which are carried up the chimney when the fire is disturbed, form the soot which accumulates inside the chimney. Soot is very easily combustible if made hot enough, so that a chimney well lined with soot could easily catch fire. It then produces a dense white smoke from the chimney which is mainly evaporated and recondensed unburnt soot. Because it condenses in the air and not onto the walls of the chimney, it is in the form of very fine particles which, looked at under a microscope, are seen to have a spongy texture.

Smoke particles are very good absorbers of certain gases, namely those which have large molecules and are easily liquefied. This property of spongy carbonaceous material is used in gas masks to remove poisonous gases from the air. In particular, smoke absorbs much of the sulphur dioxide that is present in flue gases and in a moist atmosphere it quickly converts into sulphuric acid.

In smog, many of the smoke particles were of such a size that unlike larger particles that are trapped in the nose and upper bronchial tubes, and unlike the smaller particles that were breathed in and out, they were caught in the smallest tubes of the lungs, and even in the alveoli, which are the minute air chambers on the surface of which oxygen passes into the blood and carbon dioxide is taken from the blood. Thus caught, these acid-laden particles aggravated every bronchial disorder including common colds, chronic and transitory bronchitis, influenza, pneumonia, tuberculosis, and cancer of the lung. Thus weakened in its function, the lung required the heart to pump blood through it at a greater rate, and in cases where this was beyond the capacity of the heart the wretched person lay prostrate, barely able to breathe enough to remain alive, speechless, and breathless in attempting even to stand up.

Meanwhile, healthy non-smokers coughed a little, spat a little, blew their noses, and wondered what all the trouble was about.

The increased carbon dioxide content of some smogs caused temporary intensified breathing of cold air and produced pneumonia in some old or very young people. American visitors to London lost their voices and fell victims to bronchial infections to which they had no immunity.

There were some features of the early London smogs which will only be explained speculatively. They produced a yellowish-green colour in the afternoon, when the smoke rather than a natural wet fog was the dominant component, and this hung in a deep layer over the city without reducing the ground level visibility enough to stop the traffic. The famous 1952 smog, which was the disastrous climax to a whole series, was a very wet one. To walk down the street meant that one's eyebrows, hair, and clothing became covered in dirty wet droplets: the most effective way to become blackened was to ride through it on a bicycle.

As a result of the smog, sulphur dioxide acquired a bad name as a pollutant, but we now know that if there are no smoke particles in the air it is relatively harmless to man.

Los Angeles smog is quite different. It is neither smoke nor fog. It is a sunshine haze. It is produced as a result of chemical reactions between oxides of nitrogen, hydrocarbons, and the oxygen in the air, these reactions being promoted by sunshine. The outcome is the production of oxidants, which are substances which oxidise materials without them being raised to a high temperature, i.e. without burning. Chemicals used for bleaching are oxidants, and we know how they damage materials if used excessively. Gaseous oxidants are dangerous because they cannot be avoided. They cause rubber to become brittle and so car tyres are damaged. They spoil many leafy vegetables such as spinach and beans; they stunt the growth of other plants such as tomatoes and pine trees.

One of these oxidants is ozone (O_3) which is a gas each molecule of which is composed of three atoms of oxygen. Ordinary atmospheric oxygen (O_2) is composed of two atoms, and when given the opportunity ozone gives up one atom and allows it to combine with whatever is being oxidised. Bleaching is merely the oxidation of dyes.

As explained in the previous section, the oxidants are invisible, but the main components are converted into a haze which is associated in people's minds with the smarting of the eyes although it is not the haze as such which causes it.

Tees-side smog is yet another kind of smog. Tees-side is a large industrial area around Middlesbrough in north-east England where steel and chemical and associated works abound. It is one of the world's largest centres for the manufacture of ammonia, and a certain amount of leakage is very difficult to prevent. At the same time, in such a large industrial area a great deal of fuel is consumed, and so there is inevitably a large output of sulphur dioxide.

Both these substances are emitted naturally over a large part of the earth's surface, and farmyards emit very large amounts of ammonia. They naturally combine to form small solid particles of ammonium sulphate, which is beneficial to the vegetation when it returns to the soil. In the Tees-side area however both gases are emitted on a much larger scale than is usual. At the same time Tees-side is liable to experience natural wet fogs drifting in from the North Sea, particularly in summer. The particles of these fogs are evaporated in country areas on summer afternoons and pleasant sunny days are experienced at nearby places along the coast. But the fog droplets are ideal sites for the oxidation of sulphur dioxide and its combination with ammonia, so that each fog droplet over Tees-side tended to become a droplet of ammonium sulphate solution which did not evaporate completely as natural fog did, but remained as a droplet, though smaller. The result was that as the sunshine penetrated the wet fog it was replaced by a thick white haze and considerably less sunshine than nearby. The haze made living in the neighbourhood far less pleasant. Although it was not itself thick enough to slow the traffic, it did shorten the afternoon sunny breaks and allowed the sea fog to cover the town for longer periods. It has been alleviated mainly by a reduction of the industrial leakage of ammonia. As in any modern industrial area, sulphur dioxide is still emitted in large quantities.

Fluorosis

There is much propaganda to the effect that fluorides in drinking water would help to improve the quality of our teeth. The statements in the propaganda are not incorrect; but if toothpaste and the indiscriminate sucking of sugary sweets and other foods containing sucrose were both given up and replaced by the chewing of apples and the more sensitive brushing of teeth with clean water after breakfast and before going to bed the general dental health of everyone would improve. There is probably enough fluorine in the average diet anyway, and to stop the habits which cause tooth decay would also put an end to a great deal of obesity in wealthy communities.

The point of this is to note that small amounts of fluorine are an absolute necessity for the growth of teeth, but in some circumstances animals have been allowed to graze on land near to brick or aluminium works from which compounds of fluorine are continuously emitted. Fluorine is one of the most reactive of all the elements and one of the results of this is that a far greater proportion of gaseous fluorine compounds emitted from a chimney is absorbed on the ground near the source than is true of other substances such as sulphur dioxide.

3

The consequence of this is that the vegetation may collect very large quantities of fluorine, and may grow more lusciously as a result. Cattle grazing on this suffer an excessive growth of their bones. Their joints become stiffened and their limbs and skulls become heavy, sometimes to the extent that they cannot lift their heads. At the same time their teeth become softened and stained. This disease is called fluorosis. Because of their great solubility, there is no accumulation of fluorine compounds to a harmful extent near the industrial sources mentioned and only cattle grazing exclusively on this vegetation seem to suffer. Nevertheless it is objectionable to have to exclude cattle from a rural neighbourhood because of brickworks! The chimneys of most brickworks are not as high as they should, and could, be.

Natural pollution of clean air

We have seen that the atmosphere is entirely composed of continuously re-cycled natural products; how then can we speak of natural pollution? We are not referring to volcanoes or smelly mud where vegetation is rotting in stagnant water, but to sources of solid particles which have an unfortunate effect in nature. The most important are sea salt and desert dust.

When the wind blows from the sea there is a harmful deposit of salt spray on a narrow strip of land along the coast, in which only grasses and other vegetation which can tolerate the spray will grow satisfactorily.

In some coastal areas the fields near a cliff edge are known by farmers to be poorer grazing land for cows, which yield less milk if kept there for long periods. Also it has been observed that when there is a light fall of snow it melts sooner near the cliff edge than further inland because of the salt deposit.

Trees and bushes along the coasts always show a stunted growth on the side towards the sea because it is on that side that salt is from time to time deposited on the growth points. If there is an exceptional gale in early summer growing vegetation, which normally is not subjected to salt spray because it is too far inland, may be killed as far as two miles from the coast.

Desert dust is pollution because it forms dunes which invade cultivated areas. Perhaps less obvious is the effect of desert dust on world climate. Dust reflects sunshine and radiates heat into space at night with the consequence that it tends to lower the temperature of the whole atmosphere. A very rough guess might be that if there were no desert dust the average temperature of the atmosphere might be as much as $1°C$ higher than it now is. But since that heat would not be uniformly distributed the most important consequences of not having the dust would be in the changes in cloudiness of certain selected parts of the world which we can only guess about at present. Thus, the consequences might be an increase in rain in France and a decrease in Iceland rather than any temperature change of any direct importance.

Desert dust has this influence because it is produced in areas where there is little rain. The washing-out mechanisms are therefore ineffective and the dust haze can remain in the air for a month or so. Consequently, the area covered by desert dust is much greater than by a corresponding amount of dust produced in the world's industrial areas because these are located mainly in temperate latitudes where rain is a common occurrence at all times of year. The name 'the Red Sea' indicates that there the sky always looks red because of the dust haze. The sea itself usually appears very blue by contrast! The control and cultivation of the world's deserts presents us with a challenge which cannot fail to excite anyone who has seen a desert. There, if anywhere on earth, are man's potentialities stunted by arid nature, and this thought is an appropriate one with which to turn to our next chapter.

3 The Further Evolution of Industrial Man

'Industrialisation is the commitment of civilisation to dependence on exhaustible fuel resources.'

The evolutionary situation

Individually we feel rather powerless before the giants of industry. It seems that they determine the way in which our society must develop. Even a whole nation is unable to control the industrial growth of its neighbours, and in order to maintain its place in the economic league table it reacts to competition by emulating the others. If we are to discuss control of pollution at all, we must have some idea of the direction in which we are evolving and the place of evolution in decision-making.

Over millions of years the atmosphere was changed from one lacking oxygen and nitrogen to one in which oxygen and nitrogen were plentiful. This change was brought about by the activity of living things, and much of the life which caused this change is now extinct. It lived and died with no purpose in mind but simply because life tries to go on living. Nevertheless, a purpose was achieved even though the purpose was not conceived until later when Man arrived on the scene. By evolution through conditions which cannot be maintained indefinitely a more advanced state has been achieved. The later states depend absolutely on what was achieved earlier and cannot grow out of nothing. We stand on the shoulders of our marine and reptile ancestors. Our civilisation likewise has arrived at the present via the technology our predecessors invented, and we depend not only on the machinery we have created but also upon the knowledge which made its creation possible. Our children learn about it easily because possibilities have become factual knowledge.

The reconstruction of industrial society in the desolated parts of Europe after the war was easy by comparison with evolving it in the first place or building it up in India or South America. The people had the knowledge and technical skills and there existed industrial societies elsewhere from which the machinery could be brought in and used immediately. In our rapidly evolving society those parts which were thus re-equipped even had an advantage twenty years later over those where wartime destruction had been slight.

It is also evident in the highly competitive world of commerce that as soon as an invention is known to exist it is fairly easy for someone else to invent it. The knowledge that something has been done makes the doing of it easier, so that a man from an advanced civilisation could do much better than an un-educated savage if they were both placed in equal circumstances even though at birth they might have been potentially equal. The advantages of education do not lie primarily in training for conformity within an existing society but in stimulating the imagination to make more of the opportunities which are made known through the education.

Inventions of technology are not the only inventions which upset the old routines of society: inventions of ideas are un-protected by patent law and erode the old social structures. Heresies can quickly become widely accepted. We are always in a state of ever more potent intellectual ferment and new freedoms of thought are won almost every day.

Among newly formulated bodies of ideas those we are con-cerned with here are aimed at the replacement of our in-dustrially oriented society with a new pattern of existence in which material economic advance is not a primary objective. The search for such an ideal is prompted by the obvious fact that the world is becoming very full of 'Industrial Revolution Man' and he is growing more rapidly than ever before. The complete occupation of the world will very soon be an accomplished fact.

From then onwards our objectives must necessarily change. Up to now there has always been room for new manœuvre: new land has always been available for exploitation; there has been no immediate prospect of a shortage of anything; man's power over nature was increasing so that, with the ever more plentiful resources of fuel, it seemed that any engineering enterprise could be undertaken if enough people thought it desirable.

Power is in the hands of the economic giants; in their interests, rather than in the interests of ordinary people, the main evolutionary decisions are made. Capital moves relatively freely: in consequence, those enterprises are embarked upon which have the best prospective dividends in the short term. Thus, a dividend of 10 per cent gives nothing but profit without effort after about ten years, and who can resist that in favour of something with no cash returns, especially if it will take a decade to complete, such as the landscaping of land destroyed a century ago by mining operations?

In such an atmosphere, governments have to take control of an ever-increasing fraction of the total expenditure in order to achieve social purposes which do not give the profit demanded by capital. Nevertheless we have scarcely begun a deliberate evolution towards a different kind of economy. Not only do countries compete economically, but within a country the poorer part of the population demands its fair share of the wealth available. This means producing more of the kind of things we have now: meanwhile the better-off people are acquiring new possessions and there seems to be no end to the scramble and exacerbation of problems. The momentum of the system grows and its direction can only be altered by an ever-increasing turnover and consumption of fuel.

In the past the vastness of the world has digested the pollution of evolving man, but we are now at the moment when it is failing to keep pace. Without being able to help ourselves we run ever faster along a course where we can see less distance ahead each successive year. Soon our momentum will carry us into disasters which become unavoidable because of the short warning we allow ourselves and our lack of room for manœuvre.

One reaction, common among American youth, is hippie-ism, communalism, and other philosophies which opt out of the economic competition. As economic growth becomes malignant, revulsion at the excesses of man – wealthier than he has ever been before, urgently seeking more, far beyond the point of satiation – makes thoughtful people decline to participate any longer. Unfortunately, the only result is that they opt out of responsibility for the future unless they believe that their example will soon carry the majority with them. That would be possible in a truly democratic society if those with power surrendered it because it was the obvious wish of the majority; in fact they cling to power, and it is inevitable that they should do so. The making of new laws is necessary, and these new laws must be such as to effect a healthy transformation of objectives even though society's existence will still be based on industry. Already it is impossible for the world to evolve towards hedonism without the maintenance of our industry; there will be no public acceptance of a decline in wealth except in the face of extreme alternatives such as war, famine, or anarchy. It is the task of intelligent people to prepare the way for the change and put new laws into operation before the inexorable compulsion of disaster becomes the driving force. We need to spend great effort not only in creating a vision of the new social and political order into which the world must move as it becomes full, but also in planning in detail the steps by which the evolution can be effected.

Evolution before our eyes

The blight of pollution upon our centres of industry was accepted because of the wealth and power it gave, and the opportunity which the industry provided for future advance. Undoubtedly much of our industry could have been more efficient, and the legacy of damaged land together with the experience of squalor and waste of resources could have been avoided, if provision had been made at the time to avoid them. But avoidance was not seen as either necessary or desirable. Furthermore, it will not be easy to be wise after those events unless the pressures which made our grandfathers destroy their beautiful land are removed from us. They did not think they were destroying it; they made it powerful and wealthy, and in it culture was both live and spontaneous.

We see before our eyes, at the present time, examples of evolution through a period of strain into a new civilisation not yet fully envisioned by those who are leading the change. Although imperialism by nations is at an end, exploitation of territory is as rife as ever. California has been pillaged by man as no other land in history. It has been taken by force and developed for money alone. It has undoubtedly produced great wealth not only for itself but for the world, but now it begins to pay the price on a scale never before seen. There have been depressed areas in Britain and other European countries; likewise coal-mining areas of Appalachia have been left desolate, unemployed, and ignored while the United States goes from strength to strength: but these are examples of the callousness of the economic system, and political decisions can easily rehabilitate them. Not so California. It is committed to an automobile civilisation which produces smog and destroys its vegetation. It is creating a refuse problem, a water supply problem, and many other difficulties which are not caused by economic backwardness but by the destruction and excessive use of its own environment. Californians have already sprayed so much insecticide that if they stopped now it would be several years before the height of the unintended damage would be reached. Even their economic supremacy cannot save them. They can save themselves, but only by political decisions extreme by any standards of the past. Their development has not been undertaken with a consciousness of any goal beyond the present increase of wealth. A valid and strong hope for the future lies in the fact that change is normal for them, so rapid has been their advance; a deliberate change will therefore be easier than for many communities. In spite of the awful environmental damage they have great wealth and, relative to other communities therefore, more room for manœuvre.

The position of the newly developing world is different. In Chile's capital, Santiago, the sunshine smog of the Los Angeles type is as awful as in Los Angeles itself, but no prevention of it is attempted. In South Wales the industrial town of Swansea killed its environment in the nineteenth century by its shocking emissions from the smelting industries, but it produced great prosperity for some. Santiago is carrying the Chileans into a new state of wealth in a similarly painful way. Because of the economics of the situation they can only afford it that way, and they accept it. However, the advanced countries must take much responsibility for this inability of modern man to learn from the mistakes of others; to a very large extent they are the providers of the means of economic advance of the newly developing countries, and this role is taken mainly for economic reasons. They therefore use Victorian economic criteria: the smelters of Zambia are only cleaner than the smelters of South Wales of ninety years ago because of technical advances, and not through any deliberate concern for the Zambian environment.

Under the present economic arrangements, the spoliation of land in the developing countries is permitted by all because it serves an economic purpose: if Man were to consider himself one we would learn from the past and prevent the destruction of the environment now, instead of waiting until it is so spoiled that an outcry of conservationism makes us slow down the ravages.

The disaster in the developing countries is greater in human than environmental terms. Because the industrial development in the poor countries is not home based but comes ready made from outside, it is unlike it ever was in Britain or the USA. We had to evolve painfully through social changes: the unions grew up and the exploitation of labour was gradually ameliorated. Probably much misery among employees will be avoided in the future, and to this extent bad social practices will not be exported with the machinery sent to newly developing lands. Nevertheless, the 'magic' of the old industrialised world creates unemployment on a scale far beyond the worst ever experienced, even in Germany at the height of the great economic depression of the late 1920s. The unemployed are, moreover, nearly unemployable, and become more so because imported 'magic' always advances more rapidly than local education. The development of the under-developed world creates not the slavery of the old imperialisms, but human refuse on a scale never before contemplated. We do not order people about, we simply cast them aside; we unintentionally make them unnecessary in trying to make them rich – but unnecessary for whom or for what? The presumption – that because purposes are created by man, the purposes he creates are not open to challenge – is the supreme folly of the Industrial Revolution in the context of a full world.

Not long hence we must live in a full world – a lovely world, for as we saw earlier, this world possesses a variety of beauty unsurpassed elsewhere in the solar system. The criteria and purposes which we have formulated for human existence have soon to be changed.

In advocating this change, we must not waste our intellectual effort in castigating human leaders of the recent past. They are the only past we shall ever have, and they have given us evolutionary opportunities for which we should be grateful. We need to learn to care for our planet, and not treat it as an infinite reservoir of wealth. We seem to try always to satisfy while our philosophy teaches us never to be satisfied. We have a lovable, sweet and living environment in which we are only one of the forms of life and to whose sweetness we are the greatest threat.

Perhaps we shall be able to learn best from regions of the world where the evolution is taking place before our eyes. In Britain, the deliberate and socially compassionate running down of the coal industry is an example of how the old should be remodelled. Into the coal-mining areas we have introduced new industries to prevent social degeneration such as has occurred in Appalachia. Although this has been fairly easy because the country has been very prosperous in other fields, we very nearly failed. The great depression of the thirties took us to the edge of disaster.

In Bohemia, the greatest industrial concentration of the whole of Czechoslovakia is undergoing an evolution far more rapid and dramatic. It is a microcosm of Europe in which many of the lessons will be acted out on a larger scale later on. The brown coal (lignite) lies in a very thick seam not far below the ground and with modern machinery it is now possible to obtain it by opencast mining. The seam is so thick that ordinary deep-mining methods are impracticable. Since the ground would have to be greatly disturbed by subsidence, it has been decided to work over the whole coal field in a period of twenty to thirty years during which time the town of Most is being completely rebuilt nearby and the old town completely demolished except for the church which, though not particularly old, is unique in design: it will be moved bodily on rails a distance of about a mile.

The arguments involved in the decision to extract the coal, which will provide the basis for the whole operation, are very complicated and it is worth examining some of them. The cost of this unusually deep opencast mining is such that it would be cheaper to import coal of the same type from East Germany and pension off all the Czech miners, but such an account-book conclusion is quite unacceptable because a society with one of its chief activities brought to an end degenerates in many other ways. It is important to be proud of one's mastery of one's own life and to make choices not dominated by cost calculations. Costs may be prohibitive sometimes, but they should only be used to discriminate between two similar alternatives, never between quite different ones.

Since the coal will be exhausted within a generation the new life of the area must be thought out. Power stations burning lignite will be replaced by nuclear or oil-fired stations, and chemical industries now using the coal as a fuel or raw material will turn to other sources. The mining industry itself will be brought to an end, and as the excavations proceed a complete re-landscaping of the area will be undertaken. This includes the emptying of several lakes and the creation of new ones.

Obviously there are many great issues which have to be decided as this highly developed community changes its way of life. The use of the coal is making possible the transformation to the next stage. In this sense it is a microcosm of 'Industrial Revolution Man' as a whole, for the world's fossil fuel resources will be almost exhausted in a few more generations.

The decision to embark on this evolutionary step has many side issues touching upon air pollution. Lignite contains about 4 per cent of sulphur, and a much greater quantity of ash than hard black coals. This means that the question of sulphur extraction for commercial use has to be considered. The main advantage of the removal of sulphur from fuel is that industrial countries such as Britain and Czechoslovakia import sulphur for industrial use, and this would be unnecessary if they could use the sulphur contained in their fuel. Against this, it is argued that the amount of sulphur needed is much less than is contained in the fuel used and so the extraction of it from the fuel, or from the flue gases when it is burned, could only be carried out on a small fraction of it and, therefore, could not help significantly to reduce the pollution of the atmosphere by sulphur dioxide even though this was the main reason for proposing the extraction in the first place.

On the other hand it might be possible to extract all the sulphur from this particular bed of fuel which has a high sulphur content and store it during the twenty-five-year operation for use during the following century or more. Furthermore, sulphur extracted by chemical processes is of much greater purity than raw mineral sulphur and quite new uses might be devised for it if it became plentiful and cheap. This would be a more economical use of exhaustible natural resources than simply burning the coal and emitting the sulphur, as sulphur dioxide, into the atmosphere, and is therefore commendable. It is an attractive proposition on the grounds that it creates an asset for the future and saves present-day air pollution, whereas almost everywhere we see industry exploiting the immediate opportunities with a frightening ruthlessness and leaving scars on the earth which are costly for future generations to repair. On the other hand, the advance of technology is so rapid that we cannot really be certain that what would be an asset if industry continued exactly as now, might turn out to be a white elephant if given as a legacy to succeeding generations. We have to balance the careful husbanding of resources with the possibility that their use now will enable us to advance more quickly to a stage where they are no longer necessary. Thus, coal is becoming obsolete before it is exhausted because of the advent of oil, natural gas, and nuclear energy, and perhaps we need never have worried about using it up.

Bohemia is surrounded by mountains on all sides, and is subject to periods when air remains stagnant above the ground much more often than in a windy country like Britain. Because the lignite contains more than twice as much sulphur as British coal, the problem of sulphur dioxide air pollution is quite often serious. The gases from large power stations drift with very little dilution above the height at which they were emitted, but below the mountain tops, until they impinge on the hillsides and damage the pine trees which are an important part of the economy. For this reason the possible removal of the sulphur from the fuel has been urgently considered.

Unfortunately, no satisfactory method of sulphur removal has yet been invented, and one suggestion was that the exploitation of the coal should be delayed until sulphur removal was possible. Since that would be at an unknown date in the future, the whole community of Most would continue to live in a town under suspended death sentence, without any incentive to undertake long-term urban improvements. It has also been suggested that the coal ought to be left as a resource for the chemical industry which is situated in the area. This is an attractive argument for everyone except those living in the area with the threat of mining of the coal at some indefinite future date hanging over them.

The issues have been greatly simplified in this very brief account but it is possible to see that probably the decision to go ahead with the exploitation of the coalfield now was made by men anxious to progress, to be enterprising, to be active, and to be themselves and make their own choices, groping their own way into the unknown future, placing upon themselves great challenges as their whole society is transformed. Whatever else may happen, Most is being changed from a small country town that became a coal-mining centre about half a century ago into the most modern planned living centre of the country. Most will be experimenting in many new aspects of living, whereas had the great decision been otherwise the coal would have been abandoned and Most would surely have become depressed as the enterprises which had begun to bring it twentieth-century wealth were run down.

One interesting by-product of the evolution of Most is the development of an efficient and cheap tramway system between the old and new towns. This experience is bound to affect future attitudes to public transport.

Mixed into the argument was the problem of the dust emitted from the chimneys of the great power stations burning the lignite. It is very difficult to eliminate such large amounts as are produced by burning pulverised lignite and, odd as it seems, the deposition of the dust on the countryside appears to be beneficial for agriculture. Incidentally, waste heat is used for glasshouses in which horticultural produce is grown. There are many incidental advantages to be obtained out of any situation, provided it is complicated enough!

Although the dust may help agriculture, largely because it carries with it sulphur compounds, the effect of dust in the air is to reflect sunshine and radiate heat, causing the air to become cooler and more often stagnant. It will be recalled that stagnant conditions are those in which the sulphur dioxide from the power stations is not adequately diluted and causes damage to trees on the mountains.

The place of conservation

No issue is simple: our history is evolutionary, and resources provide steps up the evolutionary ladder which, once trodden, are no longer needed. The only justification for their once-for-all use is that we step permanently to a more advanced civilisation. The car is such a step, as was slavery. If we come to see a step in our evolution as shameful, we must learn the lesson and never return to it. If the step was hard we can be proud. We are surely led to wonder whether there is anything else in the life of 'Industrial Revolution Man' than steps to a better life. At this moment in history we are compelled by the fullness of the world to contemplate human evolution without dependence on geographical and material expansion. We hope for advance in quality, but the tonnage of consumption per head has reached an excessively high level in many rich countries. Indeed the level is such that the world could not sustain the pollution if everyone in the world made as much as the average citizen in the most advanced countries.

We therefore need as a major priority to find means to control pollution in the evolutionary advances that lie immediately ahead of us.

There has always been a minority of the intellectual aristocracy of any community who have scorned the crudity of technological advance, even though their own easy life was completely sustained by it. Their theme has been, 'we now have enough, because I have enough', and it does not seem to have occurred to them that the majority of the population still want a great deal more than they have got. In a sense they are advocating the old theme that pervades all ancient stories of magicians and much modern science fiction, that the risk of the technology taking control can be avoided by 'destroying the formula'. It used to be fashionable to say, 'our problems would be solved by the abolition of the internal combustion engine'. Not only is this a drab cliché, it is an abdication of responsibility and an avoidance through fear of the exciting possibilities of the universe, and it cannot possibly ever become acceptable to the more enterprising members of any community. We see it practised in some monastic communities where full access to knowledge of the world outside is forbidden.

In looking to the future, therefore, we cannot be mere conservationists. We must preserve antiques and prevent the extinction of species such as the cowslip and the tiger, but conservationism cannot provide a distant vision of the future towards which we are evolving. To obtain this vision we need to explore diligently even such mundane matters as methods of taxation which may be as potent in directing our technological effort as any great scientific discovery.

The distant vision

It is not proposed under this heading to prescribe a Utopia, nor to investigate what kind of technology man might invent to support his civilisation in a permanently crowded world. We may only note some of the necessities under which man will operate, and try to lengthen the stretch of time at which we look.

As the pace of technological evolution and population growth have increased, instead of looking further ahead, man has on the whole shortened his vision, and in this lie the most dangerous seeds of catastrophe. We have made life so exciting and different every year that we have no need of the life after death to keep us from pessimism. The more power we have at our disposal in the form of cheap fuel, the easier it is to find new sources of power, and there appears to be no restraint upon the competitive development of any natural resource that a man can lay his hands on. Surely the moment has come to lay our hands upon the stores of wealth we find in the earth's crust in a different sense, and dedicate them to our successors. We know that man must ultimately arrive at a state of civilisation in which the world is cared for in every detail and the tonnage of our turnover does not increase. In this light the rush to spend and spend, as if that were the ultimate fulfilment, cannot be encouraged.

In order to come to terms sooner, rather than later, with the new régime which must ultimately prevail, we need to envisage it much more clearly and make this vision much more an essential ingredient of our outlook.

In past ages, when technological invention and geographical exploration were slow to advance there was a pervasive fatalism which was faced bravely even by the greatest of men. Julius Caesar, the engineer of an empire which spread a revolution in environmental husbandry as wonderful in its day as the Industrial Revolution itself, was nevertheless a man of petty intrigue in high places; so, too, was Elizabeth I who reigned over Shakespeare who, in turn, found in Caesar humanity of a kind still seemingly dominant in his own age. Although we are separated by only a quarter of the time between Shakespeare and his hero, our evolution is utterly dominated by the discoveries of oilfields and the international amalgamation of industries each consuming more energy in an hour than Caesar had at his disposal in his whole empire in a year.

Any change that is made in the future will not be of this kind. The devolution of empires in this century is a symptom of the end of the era of expansion; and if the pessimists who forecast that we shall use up the majority of our oil reserves by the year 2000 turn out to be wrong and that the correct answer should be the year 2100, in terms of the span since Shakespeare the difference is trivial. And as the span since Shakespeare is a small fraction of the time since Caesar, so these two millennia are as nothing to the time taken by life to transform the Earth's ocean, crust and atmosphere since the pre-Cambrian era into the wonderful environment we now enjoy.

In that environment, created by geological and biological evolution, are the oil and mineral deposits, concentrated against all the dominant tendencies of nature. Now we are consuming them and converting them into mixtures of substances from which it will be much more costly, in terms of time and fuel, to extract them later when future generations want them: the iron of a sunken ship rusting in the sea, the sulphur spread throughout the world as we burn it in our fuel, the cellulose in the waste paper we incinerate destroyed to become carbon dioxide and water. The production of these separated and synthesised materials will have to depend either on the slow geological processes of the past such as the separation of salt and sand by the weather or upon the biological growth of trees or animal deposits, or upon the extensive use of nuclear energy.

The only escape that we can envisage today from a highly sophisticated but essentially peasant* existence in the future is through nuclear or tidal energy. Otherwise our descendants will be completely dependent upon solar energy, the annual cycle, and the weather, and the cultivation of the deserts will be impossible.

But even with nuclear energy in abundance the cultivation of the deserts will necessarily be slow. It will require a gradual experimentation and development: it will not be possible to embark on it except by a slow advance at the edges and a consolidation of every small advance by a race of men who learn new skills and invent new machines and substances to meet all the problems as yet unimagined which will arise as we disturb the state which Nature arrived at over thousands of years of evolution. Even if we were to achieve an invention as spectacular as Nature's discovery of chlorophyll, which in thousands of millennia of the distant past replaced the carbon and hydrogen compounds in the atmosphere by oxygen and nitrogen, it would still be a very long time before any significant change could be brought about.

What do we mean by a very long time? The answer to this question reveals the truth about the change in outlook required by man. The progress of the Industrial Revolution has made us ever more impatient: success and new wealth come quickly, and we hate to be left behind by others. The most obvious numerical measure of what is happening to us is the high interest rate that is now normal. It signifies that more wealth can very quickly be obtained and that anything old that does not have value as an antique soon becomes worthless. We cannot store wealth: we live with galloping inflation. In the

---

* A peasant existence is one in which succeeding generations are bound by more or less the same physical circumstances, and a person's life is not dominated by the need to save and invest for advance to a quite new kind of existence. Clearly we do not think of our descendants as reverting to the old kind of peasant life, but the changes will not be concerned with more food, more health, more personal possessions, but with the problem of providing more individual opportunity for creativity.

4

next decade we shall create more wealth than the total we now possess, so how can we plan or envisage a life beyond that time? We feel bound to grab our share or be left behind powerless at the end of ten years. Grandchildren have become an irrelevance, heresies are rife, and unless we take a firm grip upon ourselves our culture will degenerate into anarchy through lack of permanence. The democratic society is in danger of squabbling its way into slavery under the growing power of business over technology and the computerised dehumanisation of our social organisation.

In the ultimate peasant existence that would emerge, our cities would be like anthills. As one human falls he would be cleaned away like a wrecked car and the machine would grind on.

On the other hand if we can extend our vision and take control of the economic forces which at present direct all our technological development we will easily be able to preserve our culture. The compelling logic of the market has served us well, but it must be replaced by new criteria in a full world.

The immediate need

The exhaustion of natural resources is a serious and dangerous possibility. Nevertheless, by present-day standards it is very distant indeed, and under our noses are problems requiring immediate decision. Our next task is to see how to determine priorities for action. This is as much a technical as a moral or political matter, and we now turn to some of the details.

4 Local Air Pollution

Effects of pollution on local climate

At Tyrone and Lock Haven on the northern flanks of Bald Eagle Mountain, a long Appalachian ridge in Western Pennsylvania, there are wood pulp mills which emit several kinds of air pollution. Apart from the ordinary products resulting from the burning of fuel there is a large amount of water vapour produced by the drying of the pulp and a sodium sulphite aerosol. The consequence in fairly calm weather is that the air in the long valley between the mountain and the next ridge which lies parallel to it is filled with what looks like a white smoke haze. The sulphite haze particles not only reflect a large fraction of the sunshine which prevents the ground from becoming as warm as in the neighbouring valleys, but they also radiate heat into space and have the effect of cooling the air more than if it were clean. By the well-known greenhouse effect the upper layers of the atmosphere prevent clear air near the ground from cooling quickly by radiation; but the particles, which radiate energy in wavelengths not absorbed by the clear air above, cool the air in which they are suspended to such an extent that water begins to condense on them. They grow in size and their radiative effect is correspondingly increased.

The conversion of the haze into a wet fog is encouraged by the hygroscopic nature of the particles and by the high humidity of the air containing them caused by the emission of water vapour from the same factories. Very often a fog as dense as ordinary natural cloud is produced and this is more persistent when the sun shines upon it than a haze.

Sunshine does not evaporate fog directly: fog is 'burnt up' by sunshine only when enough scattered sunshine penetrates through the fog to the ground, which it warms and which in turn warms the air above it. The fog is then gradually evaporated. Cloud particles absorb a negligible amount of the sunshine falling on them, whereas the ground absorbs about half of it. This figure of a half is very rough: the precise fraction depends on the colour of the ground and other physical properties, and varies between 0·9 and 0·1 among different common surfaces.

The direct consequence of this is that on many occasions the air in the valley on the north side of Bald Eagle Mountain is cloudy and cold while the sun shines brightly on the air in the valley to the south.

We have already mentioned in chapter 3 that the air in north-west Bohemia often becomes stagnant because of the surrounding mountains. This happens about thirty-five days in the year, and the stagnation of the air is intensified by the presence of pollution in it. Most of these days are in the winter and the stagnant smog fills the region like a lake. In summer, the mountainsides above the smog become warm in the sunshine and this causes the air to be mixed at the 'shore' of the lake of smog; but, in winter, the mountains are covered with snow which reflects sunshine as efficiently as clouds and also loses heat through clear air as effectively as dark-coloured ground. As a consequence, the local climate is dominated at certain times of the year by the effects of pollution. The effect is always adverse.

A rather different situation obtains at Fairbanks, Alaska, where the days are only an hour or two long in winter. Fairbanks lies in a valley at 65°N and its power stations burn fuel containing hydrogen. Consequently, there is water vapour in the combustion products and, instead of remaining in vapour form as is the case in most of the world, it first condenses like an aircraft trail into a cloud of liquid water droplets and then freezes into ice crystals because of the low temperature. The cloud persists, and fills the whole valley, darkening it even at midday and causing a stagnation of the air. One of the consequences is that the air is more moist than it would otherwise be and car exhaust fills the streets with more condensed cloud. Fog is added to the darkness, with no relief at midday.

The advantages of the climate of large cities

The combination of pollution and a situation in which fog would naturally occur is the cause of the worst smogs. But the urban climate is not affected only by the pollution. The continuous generation of heat in houses and by traffic keeps the temperature of the air in the streets higher than it would otherwise be, and this tends to reduce the frequency of fogs. Furthermore, the rain that falls on a city is drained away so that the streets and roofs are dry most of the time.

Green fields may seem dry, but the vegetation can only remain cool and green in sunshine if it can continuously draw water up through its roots and evaporate it through its leaves. It keeps cool in the same way as we do by evaporating sweat. If there is no water at the surface, and no roots to draw it up from below, the surface becomes hot in sunshine, so much so that it may be too hot to touch. Car bodies, dry sand, and house roofs become much hotter by day than any objects of the natural countryside, and the air becomes drier because there is no evaporation. Consequently, by night, the air is warmer and less moist and fog occurs less often than it used to before the town was built. Mist patches are common in fields and large grassy parks, but do not occur in built-up areas.

A clean urban climate is healthier to live in. The advantages of the countryside derive more from the kind of activity that the human body is engaged in than from the chemical nature of the air. This is not to say that an urban climate is always better than a rural one, but that it lies within our power to make it better for living in, if we can get rid of objectionable air pollution.

The effects of local climate on pollution

Just as an unfortunate geographical situation makes pollution have a bad effect on local climate, the same circumstances can produce a pollution problem because of the already existing climate.

Inland from Los Angeles and surrounding Santiago in Chile, there are mountains large enough to prevent the air over the city being carried far inland by sea breezes. The air is kept stagnant or slowly circulating between the coastline and the high mountains, and in this situation a sunshine smog is produced. There are many cities where this will never happen because there are wide plains inland and the sea breeze continually carries fresh air across the coastline. For example Karachi, Perth, Casablanca, Cairo, and Buenos Aires will never suffer a major car-exhaust sunshine smog, although many inland towns shut in by mountains may, and in some cases already do. Among these are Athens, Mexico City, Tashkent, Tehran, Lyons and Ankara. Cities large enough to produce higher pollution levels because of their size are, of course, liable to urban sunshine smog. Paris, New York, Tokyo, and many sprawling cities of the United States already have this problem from time to time.

Much of the dense urban development at Wellington (NZ), and more so at Lower Hutt near by, is mainly on low-lying land which has either risen naturally in very recent times or has been claimed from the sea during the last hundred years. This is surrounded by steeply sloping mountains, many of which are unsuitable for quick urban growth because of their continued geological evolution and frequent landslips. In calm weather, therefore, the air pollution is trapped within Port Nicholson. The pollution often impinges on the hillsides, but does not yet reach serious proportions. It will, however, almost certainly get worse before steps are taken to make it better.

Although the mountains are on only one side of the cement industries on the coast of Lebanon these are, in effect, shut in by them at times of calm weather in the eastern Mediterranean. Flow of air over the mountains is inhibited in the winter when they become snowcapped, and the plumes from the factory chimneys at the coast impinge on the rich fruit-growing fields which slope up steeply inland (towards where the Cedars of Lebanon used to grow). To some extent the trade provided by olives and citrus fruit is being sacrificed for the greater profits from cement manufacture.

Steel works are very much the same the world over, but the effects of their pollution depend considerably on the local climate. At Consett, near Durham, the works is on a hill and its dust is carried away very effectively by the wind so that it has little effect on the neighbourhood at ground level. By contrast, the red dust, characteristic of high quality steel making, became so dense at Linz in Austria that it has been necessary to install expensive filters to capture the particles before emission to the air. Linz lies deep in the valley of the Danube and, being subject to a continental climate, has longer periods of light winds than are experienced in maritime countries. Its climate is worse for steel making than the Rother valley at Sheffield, Ebbw Vale, or Scunthorpe (which stands on a ridge). Pittsburgh, on the other hand, experiences a continental climate but lies in a much more open valley than Linz. At Linz, it is said that the recovery of the red fume repays its cost: at Pittsburgh, the great improvement in air quality is mainly due to the replacement of wasteful old furnaces and coal house-heating systems which produce dense smoke by new, more efficient, and more economical ones. It is characteristic of many improvements in air quality that they have paid for themselves in this way.

The Hudson valley extending two hundred miles northwards from New York to Albany is very suitable for many kinds of industrial development. In particular, the plentiful supply of water makes it suitable for the siting of power stations which need a medium for waste heat. Already the river itself is receiving waste heat up to almost the maximum allowable. With further rises in temperature there could be serious health problems caused by the inflow of urban sewage, and the almost complete deoxygenation of the water, especially in summer. Naturally, therefore, engineers turn to the possibility of using ordinary wet cooling towers of the kind familiar in Britain. Unfortunately the winter is so cold in New York State that not only would the use of many cooling towers fill the Hudson Valley with fog, but the fog could often become supercooled. If it were as cold as at Fairbanks and the fog particles froze and became ice crystals the situation might be tolerable, but a supercooled fog (i.e. unfrozen, but at a temperature below freezing) creates impossible conditions for transport. All the fog droplets which make contact with solid bodies freeze to them, covering them with a layer of clear ice. Roads would be coated with slippery ice; overhead wires would become sheathed in ice and often overloaded to breaking point; bridges would be particularly exposed because of the greater flow of air past them than past objects closer to the ground, and they would become quite impassable. Switchgear at the power stations would become inoperable. Yet the power stations would be technologically indistinguishable from those operated without inconvenience of any kind all over Britain, and adjacent parts of Europe: the local climate makes all the difference.

The local climate may interact with pollution to produce many kinds of problem peculiar to a locality and often they are insoluble. Any attempt to solve them by legislation applied to the whole country would be mistaken. A car travelling across California only contributes to a pollution problem when it enters the low lying areas such as around Los Angeles. A steel works or a power station may be a problem in one place but not in another, and an efficient use of available resources requires us to apply the expense of pollution abatement where it can produce the best return for the effort.

Meteorological control

It it is only in some places that certain kinds of pollution become a problem, it is equally only in certain kinds of weather that this happens. An idea, much favoured at one time, is that if it is expensive to eliminate air pollution we should only do it when the weather is such as to make the pollution harmful. It sounds very fine merely to turn a switch and change the fuel of a power station from oil to natural gas, thereby stopping the emission of sulphur dioxide during periods of air stagnation, or to pass waste heat into a river in cold winter weather and into cooling towers at other times, but the cost of constructing and maintaining in readiness two sets of equipment is usually so large as to make the idea impracticable.

Among other considerations are that land, which is very valuable in highly industrialised areas, would often have to be occupied by the unused part of the works, or kept for storage of alternative fuels, and the effort in most cases is best applied to the elimination of the problem altogether by other means. For example, if we could devise a dry cooling tower less expensive to construct and more efficient in operation than those that have been built so far, not only would the Hudson Valley power station problem be solved, but the need to site a power station by a river in the first place would be eliminated.

In the case of the Linz equipment for the prevention of the red dust emissions it is claimed that the recovery of material pays for the cost of the operation, yet it was the decision to eliminate the pollution and not the prospect of profit which stimulated the design and construction of the equipment. In a similar way, the elimination of smoke from household fires in Britain has made us use more efficient stoves and we get better value for money spent on fuel all the time as well as cleaner air on the days when pollution might have been bad. (Actually, in the case of domestic smoke, the reduction in street air pollution is very significant in average weather, and it is only on a very few days in the year, such as when there are gales, that the improvement in air quality is not appreciable.)

It is, therefore, not generally worth while thinking of selective part-time elimination of pollution as a solution of the problem. It is very worth while eliminating domestic smoke in cities all the time, first in the centre and ultimately everywhere. But it is very worth while eliminating car exhaust first in places where it produces the worst effects, instead of trying to make the improvement everywhere at the same time. Cars which never go into towns do not need clean exhaust until it can be achieved easily and cheaply. Cars present a special problem because any car might theoretically go anywhere. But a scheme whereby only those with certified clean exhausts could enter certain restricted zones in cities might enable us to gain experience in the development of clean exhausts. We would only clean up the exhaust of all cars when a cheap effective method was available. The alternative is to spend a great deal of money cleaning air where no harm is done and allowing cars with obnoxious exhausts into cities, where harm is done, during the long transition period which would be necessary.

Perhaps the greatest objection to meteorological control is that it can be very costly in a highly developed industrial society. Modern industry is designed on a continuous-flow basis. It operates day and night, every day of the week, because productivity is much greater that way. Some parts of it, such as the blast furnaces of steel works or units of a petrochemical industry (making plastics among other products), may take several days to stop or start. It is the height of stupidity to wait until the weather has caused a dangerous level of pollution before ordering a shut-down, because by the time it has been effected the weather will probably have changed and the threat passed. Because forecasts of the weather for more than a day or two ahead cannot be certain enough, a shut-down in the expectation of smog weather two or more days hence could cause needless dislocation of industry. The wind might continue to blow contrary to expectation, or it might fall calm when no shut-down had been ordered in time.

Whatever one may say about the morals of financial profit as a motivation in modern society, there is no doubt that a shut-down of an expensive and enormous modern industrial complex is very costly and frustrating for the whole community. The starting up and shutting down are also likely to cause periods of worse pollution emission. If there is a likelihood of occasional danger through air pollution, the construction of the complex in that form and in that place should not have been allowed in the first place. There are legal and even constitutional difficulties in prohibiting for short periods, occasionally and at short notice, an activity which is permitted the rest of the time. There is an important distinction between keeping an industry going but temporarily with more expensive raw material as would happen if a power station were to change to a cleaner fuel, and stopping the operation altogether. The extreme folly of such a measure would be to prohibit the use of coal fires in houses on cold winter nights, on the grounds that it would cause an air pollution danger, when people had no alternative means of heating their houses. There would probably be a great increase in ill health among the normally healthy part of the population, whereas the health damage of urban smog of the London type can be avoided if people likely to succumb remain indoors.

If we have the means to prevent pollution being produced we should use it all the time as a civilised habit, and not think of pollution as being objectionable only when it is serious. Serious pollution is an insult to the world which has created us, which sustains us, and with which it is a sophisticated joy to engage in the pleasant exchanges of biological activity. To learn to love the world of which we are part is to hate greedy disdain for its beauty.

If that is not appreciated for the very practical aspect of life that it is for the complete human, there is another objectionable aspect of meteorological control: namely, its essentially prohibitive nature. In the first place it is inviting disrespect for the law if the law is difficult or impossible to obey without a feeling of injustice. The policy inherent in the British Clean Air Act is to equip everyone to live without making smoke and then expect everyone to behave decently and not make it. Indeed it goes further: it makes it difficult to make smoke, and everything it achieves is taken for granted as normal in an advanced society. It is not anticipated that enforcement is part of the operation: the new pattern of behaviour is instituted by creative acts, which is quite the opposite of trying to get rid of the old, dirty habits by prohibiting them, and inflicting punishment on those who do not obey.

The creation of disaster potential

The disastrous nature of natural events is not necessarily inherent. A volcano in a remote desert can be a great and glorious spectacle as far as man is concerned. But if it engulfs a town, we call it a disaster. Likewise, earthquakes and landslides may be dramatic phenomena, and only become catastrophic when they destroy buildings. We measure the severity of cyclones in the Bay of Bengal not by their wind strength but by the damage they cause and the magnitude of their tidal waves which flood cities and villages.

Thus, Man creates sites for disasters. He asks for trouble if a city is destroyed by an earthquake and he rebuilds it in the same place, especially if it has happened more than once. A city smog is not at all an untoward event as far as Nature is concerned because the air becomes clean again as soon as the wind begins to blow, and within a year all damage to vegetation is made good. Nature has no lasting memory of smog incidents, and it is solely the damage to human health and property that makes us think of them as disasters.

Man always hopes it won't happen again, at least not to him. Nowadays we can rebuild so quickly that the scar of an earthquake can be made to disappear in a very few years. We soon forget the loss of people and look like being much more concerned with overpopulation. Just as the population, though decimated, returns to the scene of the cyclone or earthquake, it continues to expose itself to smogs. In Los Angeles it is almost as if smog were the normal environment of work, the occasional invasions of clean fresh air being like weekends to be enjoyed before returning to the toil which makes men rich. The smog of Santiago is quietly borne because everyone is much more conscious of the growing affluence. Foul though living may be, most of us do not know how to avoid being drawn into the developments which promise the quickest increase in wealth. Many growth points of modern industrial states are not planned at all: in so far as one needs a plan to bring about orderly and economical construction of course plans exist but, as we become wealthier, we do not envisage life more than a decade or two into the future. If we can get our children safely into maturity, we have discharged the responsibility that society demands of us and the wonderful life our grandchildren will live is not for us to constrain; it is beyond our imagination. We live so much for today that we scarcely guard against anything in our scramble not to fall back in the league table of prosperity. In this way we create disaster potential.

Disasters are already upon us, and perhaps the most important aspect of them is that they are almost all caused by a fortuitous concourse of causes peculiar to the locality. The most famous pollution disaster of today is Lake Erie. When writing the previous paragraph I was thinking what the intelligent reader might think of the words, 'We live so much for today . . .' 'He can speak for himself!', the reader might think. 'Perhaps he is a thoughtless profit seeker, but most of us are quite ordinary citizens: not particularly anxious to climb; not afraid of others getting rich; loving our gardens and our holidays; developing comradeship at work: we do not pursue the rat race; we are trying hard to be content and find some peace, beauty, and constructive satisfying activity in our lives.' And so it is: the moralists who generalise assume that unwary acquiescence in the undesirable trends of civilisation is the same thing as deliberate pursuit of them. Politicians say, 'You can't blame the government for pollution: it is you who want all the products of a modern industrial society: it is you who cause the pollution. It is up to you to choose between wealth and pollution on the one hand and cleanliness and poverty on the other, and you always choose pollution, and you make it necessary by having too many children.'

If we deny that we make demands like those which cause the death of Lake Erie, we need to find some characteristic of the eleven million people who live around the lake which makes them worse than us. In fact their only crime is to live around Lake Erie; in other respects they are very like many times their number who do not at present make a direct contribution to a growing pollution problem.

Nevertheless, the eleven million who pollute Lake Erie produce for a much larger population many things they consume. In a sense Lake Erie is suffering the pollution produced by satisfying the so-called needs of many more than eleven million people. It is a disaster gradually created by no deliberate wish of anyone: everyone who contributed to the disaster was acting within the laws and conventions of society. The slow deterioration of the lake has been watched in helplessness because the laws and conventions are made by a much larger society than that which exists by the lake. Laws which would have protected Lake Erie from its dreadful fate would have been unacceptable in many other parts of the US and Canada. Now that Lake Ontario is well on the way to the same fate the whole weight of constitutional law operates against saving it because regulations designed to save Ontario would be contrary to the interests, for the time being at least, of those using Lake Michigan, Lake Winnipeg and many other lakes. Those using Ontario don't want to be put out of business by discriminating prohibitions.

Thus the simple fact that the countries surrounding the lake are large, and care of the lake is the interest of only a minority, make it impossible to change the laws and conventions of the larger society to prevent disaster. While Lake Erie is killed by water pollution, a large part of California is being slowly stifled by the continued extension of the sunshine smog. San Francisco sits on an earthquake site. Japan's industrial pollution kills the fish around its shores upon which her people traditionally live; the French Riviera spoils the water which attracts visitors to its coastal hotels by putting the sewage untreated into the sea. Belgium attracts industries by its freedom from anti-pollution laws: already its rivers are black and carry no fish because none of its major towns has any sewage-treatment plant. Consequently, other abuses of the rivers as dumping grounds for waste go unnoticed.

Every time a pollution danger appears it is caused by a complicated mixture of circumstances which overcome the restraints which have been adequate almost everywhere else. The growth of modern economics and the cheapness of transport have made possible very rapid local growth of a special kind. The phenomenon is not new for we saw it in the nineteenth century in Britain when the countryside of the Black Country and parts of South Wales was destroyed, and has still not recovered. The disaster potential lies in the system which is not flexible enough to enable local rules to be willingly operated. Commerce is international: even states which exert tight controls over trade across their frontiers are engaged in competition at the international level so that they, too, are subject to the same forces.

The rapidity of modern growth, the fullness of the world, and the internationalism of trade make it difficult to act to prevent disaster, even when its imminence is recognised. The causes are so special to the locality that no one living elsewhere recognises the need for a change of habits.

Disaster potential is created by the inflexible pressures of a worldwide system unrelieved in sensitive areas. A major task of civilised enlightenment is to relieve the pressures before we are forced by an ever more frequent succession of increasingly serious pollution disasters to act at last in desperation.

5 A Theory of Anti-pollution Law

'Opposition to blind progress – not blind opposition to progress.'

The liberating purpose of a theory

The spread of monotheism in the first Christian millennium was necessary to liberate humanity from superstition. It made the advance into the scientific age possible. Magic was no longer an arbitrary phenomenon by which capricious spirits could be flattered or cajoled into doing men's will, and miracles were organised in men's minds on a more orderly and universal basis. As soon as miracles became explicable in terms of a single God, inevitable logic gradually eased them out of existence.

Jesus was a rebel against the hierarchy and from his movement some people got the idea that every individual could have access to the one God on a personal basis. This very idea must have played a part in enabling individuals to entertain quite novel ideas about the physical world which contradicted the words in the books of the magicians. The struggle against superstition, idolatry, and unquestioning respect for authority that turns out to be arbitrary, has not been won for all humanity, but the enormously increased rate at which orthodoxies are challenged today is very heartening because otherwise humanity would have moved inexorably towards the environmental catastrophe which extrapolation of present trends makes inevitable.

The concepts which 'Industrial Revolution Man' teaches his children seep into his total being: they involve his every positive activity and this consolidates them. However much he deplores the extravagant rat-race of modern industrial society he deliberately takes part in it and enjoys it as much as he can. He fears being left behind with no control over his fate

55

if he opts out. Fear of magic, fear of hell, fear of missing something, fear of the obviously unknown future has driven people along courses which they choose reluctantly, but choose nevertheless because they are not liberated from other people's assertions and the apparent immutability of law.

The subject of progress and pollution of the environment is so complicated and full of contradicting advice and expressions of powerlessness that some concept is needed which will enable people to give their chaotic and disorganised support to effective action. Every flamboyant anti-pollution act that produces no long-term results holds up progress because it seems like progress while actually delaying the setting up of the organisations in society that will ensure proper control of pollution.

A grand liberating theme that will enable people to reject the assumptions and threats of the present-day power structure which causes pollution to be made is not enough in itself, although it is probably necessary. It must be backed up by well thought out programmes of action, which means a great deal of homework by everyone who intends to take part.

The grand liberating theme which is evident today is the rebellion of people against the legal system and the presumptions which keep power in the hands of the giants of commerce. People are realising that the inherent rightness of the law is a fallacy: the law is a means to keep the present power group in power; to prevent insurrection, rampant heresy and, it is believed by most people, anarchy. People more than ever before are questioning the status conferred by the law on the instruments of 'Industrial Revolution Man'. The heresy is not new; it is a recurring theme through the last two thousand years, but ordinary people are beginning to believe it is necessary, and expect to be able to do something about it on a grand scale. That is new.

Black Power and Women's Lib reject the conventional compulsions of wealth in favour of the compulsion of personal dignity. It is frequently argued that the Africans in South Africa are economically better off than in African countries with African governments and therefore it is in the Africans' best interests to live in a country like South Africa. To which there is no rational reply! The premises of an argument are never rationally based and it is they that are in dispute in this case. Why is it assumed that economic advance should be an overriding consideration in men's minds just because it has been the first priority in the minds of the men of the Industrial Revolution?

When is enough wealth enough? One must suspect that a fear lurks in the minds of those who have created wealth that if it is used for some other purpose than gaining more it must be for a debauching sensual orgy, unless the technocrats direct our energies to conquering the Moon.

In all our discussions of how to control pollution there must be a reorientation of our assumptions. The premises from which we argue must be changed. The philosophy of 'Industrial Revolution Man' has been thoroughly worked over like the theology of the great religions. Conversions are indeed made to and from these religions, but not on the basis of the philosophy, theology, or logic: they are prompted by love, fear, sickness at heart, horror at the turn of events, and so on. The logic will try to keep the machines of industrial society marching forward to new heights of progress and only a great public fear or rejection of the declared purpose of the machines can alter their course.

At present there is no unanimity of outlook among those who deplore the supremacy of industrial power. The Black Power movement has such intense beliefs in the equality of man that economic criteria are cast aside as obstacles to real individual choice. Nationalism rejects the domination of one nation by the economic power of another but clings to the concept of territorial government and absolute property rights over minerals within national frontiers: this is an unfortunate tactic because it affirms the priority of industrial wealth over people. Hippieism rejects economic power, but has no plan for the economy which supports them. Luddism rejects technological progress because it takes dignity from a man and turns him into a part of a great machine, but Luddism is a negative philosophy. And finally, those who love Nature and the beauty of the world, and hate to desecrate it with pollution and dereliction, can strike no chord of sympathy in the billions whose experience of life is hunger, grime, crime, persecution and exploitation; for them, nature worship as a vital motivation is dilettantism.

In such a chaotic scene should we then turn to the prophets of ecological doom to find a common fear to unite us all for action? Their prophecies are usually no more reliable than those of economists, and for every Jeremiah there is a technocratic optimist. In any case, pessimists are unfit to be leaders because they are not committed to exploring unceasingly for a hopeful outcome. The trouble with too much prophecy, whether of doom or a golden age, is that logically it is usually correct; but its assumptions are mostly naïve, simplistic, and do not recognise the strength of complexity.

Nevertheless, we shall certainly suffer a succession of disasters. The question to be faced is how severe they need to be before we set ourselves on a path that does not make them inevitable. We need to shed many prejudices which keep us on our present course and the political will to change can only emerge when a well-mapped alternative is evident. To recognise it as well mapped the public needs to have thought about it for several years and so, while we liberate our minds from the assumptions of the Industrial Revolution, we need also to explore diligently the consequences and implications of alternative outlooks. The purpose of this chapter is to survey some of the alternative kinds of action that are being discussed and tried at the present time.

To suggest a way out of the present impasse which is utterly heretical and contrary to accepted customs and which may never be employed may, nonetheless, serve the purpose of making what has up to now been accepted as obvious seem less convincing, and prepare the way for another equally radical alternative.

Do we need any new anti-pollution laws?

There is a school of thought according to which the evidence of the past justifies every optimism. We have suffered terrible disasters of disease, famine, earthquake, war, tyranny, slavery, and cruel exploitation of man by man and yet we have emerged wealthier, freer, and more confident in our powers. We can point to quite specific and successful action taken to control pollution. There is nothing special about pollution, it is in the same category as all the other undesirable things which humanity has managed to rid itself of and requires no special theory of politics or economics or moral philosophy.

This kind of argument can be made very attractive, particularly because it tells us that we do not require any new attitudes: man is so resilient that a way out will be found from any difficulty. It is reinforced by statements that the prophets of doom have always been proved wrong. Doom has certainly struck down many civilisations but never all of humanity. We have marvellously recovered from Hitler's war in our own time and that was worse than any pollution, was it not?

Such arguments seem to ignore the misery that has actually been suffered, and imply that if complete and absolute doom is not inevitable then there is nothing in particular to be avoided.

If humanity continues to proliferate and industrialise according to present trends starvation for the majority and gross pollution everywhere are absolutely inevitable. This is because of the inequalities between people and nations. The rich may well advance to a static level of consumption and pollution, but if they slow down because of the physical limits imposed by the size of the world they will find the world inhabited also by a much greater number of people whose misery and poverty are incompatible with education and any concept of the essential value of the individual, regardless of race or class. At present, the rich make so much pollution that if all the world lived their way there would indeed be an ecological doom at hand. Doom is not predicted by that statement: it means that the rich nations must find a way to live in which they make less mess of the world than now, in which they use less of the world's resources and make less rubbish and noxious by-products per head.

Against this it is argued that the predictions about the population explosion have been wrong in the past and that there is still plenty of room in the world for many more of us to be fed. Malthus's predictions went wrong because food production was increased by advances in technology far beyond what he had thought possible; but that does not justify taking risks with the future on the grounds that, by analogy, future advances will be equally successful in keeping starvation at bay. Everyone knows that there must be a limit to growth of population and pollution, and even if some people don't think we need to worry about it now, they do not say that harm is done by thinking about it. So let us think on.

The idea that there need be no essential change in our outlook and practices is false because it does not acknowledge the novelty of the present situation. The Industrial Revolution is less than three hundred years old. In its early stages, when shocking pollution was produced in many places and conditions of work in factories were appalling, only a small fraction of the area and population of the country was involved. It was still possible to get away from it all in an hour, on foot!

But now we are all deeply involved in a society that doubles its consumption of fuel every decade or two. It is not the way of life that has suddenly changed (although in geological and evolutionary terms a change in three centuries could be described as sudden) but the scale of the operations. Our way of life depends entirely on fuel consumption, which has grown in a new way since the Second World War. The grand scale of our operations is a very recent phenomenon indeed. Not only has the population increased but the average person uses many more materials in his life. The tonnage of earth shovelled, of ores mined, of fuel pumped, of by-products and waste disposed of by dumping, of refuse incinerated, of pages printed, the miles travelled and the buildings erected – all are increasing every day at a growing rate because of the increased power at our disposal through fuel. We use up space and resources at a hundred times the rate our grandfathers did. There has been no recognition in the economic system that people can be satisfied with enough; and from providing necessities, our social machinery has turned to the promotion of sales for its own sake. This is a new phenomenon for the population as a whole.

The social consequences in the developing countries of the rapid advance of technology in the rich ones is that the poor are increasingly made dependent upon the rich. They have no industries capable of producing, or even maintaining and bringing up to date, the machinery they import from the rich. The unemployment in the rich countries, which arises because they do not know how to organise a stable society and are obsessed by competitive advance, is nothing compared with the unemployment and social disintegration they cause by the help they give to the poor. When they supply the latest equipment to a backward country, the industrial community imagines it is helping to create wealth there; yet the most important effect is to make people useless. And when they feed starving children, keeping the economic system unchanged, they enable them to proliferate a new generation of underfed unemployables with no prospect of a dignified life. By creating people who must live without dignity, any prejudice they may already have that all people are not equal is reinforced. And then it is said that people are not equally deserving!

These issues cannot be ignored in discussing air pollution, because pollution is being introduced into the developing countries and an economic system is being built up in imitation of the system that has produced pollution in wealthy countries. At the same time a vast unemployable population is also being created as an additional burden upon the poor. They cannot even be exploited as the working classes in industrial countries were in the eighteenth and nineteenth centuries.

This all means that there is no special or political analogy between the industrialisation of the whole world and the earlier, very limited, Industrial Revolution. It is the last attempt of the centres of industrial power to maintain their power. It will intensify the growing differences between rich and poor because there is no profit in the poor. Already we, in the rich countries, create so much pollution that there would be an immediate ecological catastrophe if every person in the world made as much as we do. This is the measure of our predicament: to progress as before is to condemn the majority of humanity to a lower level of existence without any hope of equality for their descendants. If the population and industrialisation grow as at present it will make any idea of justice between man and man, between nation and nation, quite impossible.

Since there never has been justice, shall we be any worse off? The characteristic of a modern technological society is that its strength depends not upon human power to govern or lead or fight, but upon the use of fuel: differences in wealth no longer have any of the traditional justifications. In a modern society, communications are such that rich and poor cannot be kept apart and educated for their special roles. Unless we utterly reject the democratic ideal, which was the ultimate justification of the Industrial Revolution with its promise of rising wealth for all, we must reject the permanent stratification of society on a grand scale. That means learning to make less pollution than we do today in wealthy communities.

The avoidance of disaster potential

If pollution comes upon us as a series of ever more frequent minor disasters, our reaction will probably be to try to plan disaster potential out of existence. The tactics are clear: the government must control development by restrictive laws, licensing, and deliberate promotion of industrial development according to plans which keep dangerous technological products such as nuclear energy, pesticides, and so on, in their proper restricted place; emissions of air pollution in places where adverse weather is known to occur must be avoided, and economic incentives which make people take risks and create disaster potential must be controlled firmly.

Such tactics are necessary, and will to some extent certainly be employed. They are the methods which our optimists expect to use anyway. But we still need a strategy to guide the transition from an ever-increasing tonnage of consumption per head to a stabilised society.

This stabilised society is not a kind of ideal to be formulated now and then adhered to for ever; it is only stabilised in respect of population, consumption, and pollution making. In all other ways it must be expected to develop. The reason why a new strategy is required is that in order for change to occur we need room for manœuvre. Our present organisation has grown up on the assumption that there is always new land available for development and new resources to exploit. It will be necessary all the time to hold back from complete occupation of the territory available if only because unforeseen disaster potential continually arises.

Prohibitive legislation backed up by penalties

Even with a strategy the tactics of avoidance will have to be employed all the time. It seems obvious that if a practice is undesirable it ought to be prohibited, and that prohibition cannot help but reduce an offence. If stealing and murder are largely prevented by being prohibited surely the same is true of pollution nuisance!

A breakthrough into new conventions of thought is certainly needed here. We do not need to change human nature: all we need is a better understanding of why people behave as they do. The law prohibiting a practice codifies what society expects of its members; it is not the root cause of the conformity, if conformity there is. It does not require any sophistication to see that to live within limits convenient to others is desirable: even cattle and hyenas accept this. For this reason, if people can live conveniently without murdering or stealing they will do so. Most individuals who murder and steal do so only when life is too difficult otherwise, and perhaps the difficulty arises mainly through the flaunting of wealth or privilege by others and not through necessity.

We pollute because we want to keep warm and have some of the other good things available in life. We don't pollute deliberately, but as a consequence of some other desirable activity. We would willingly switch off the pollution if we could do so without giving up anything, and we will go on polluting if we find it too difficult to give something up.

To introduce a prohibition as a means of stopping pollution creates new crime, merely by definition, if it is not easy for people to live without making it. Thus, it is not practicable to prohibit cars in Los Angeles to get rid of the smog because people cannot carry on the life they are used to without cars and the city is not planned for any other kind of life. Only if some equivalent form of transport were available could it be made an offence to bring a car into the city. In that case the law could easily be enforced by having a boundary inside which no internal combustion engines were permitted. It would work only if there were a happy community inside and no rebellious mob at the gates.

Thus a prohibition makes a negligible contribution to the avoidance of pollution; it merely codifies what has become possible and desirable. For this reason a prohibition should not be introduced until it is known that people can and will comply. If a community is not convinced of the value of a prohibition they will flout it. If, for example, a law were made that no untreated sewage may be put into the river after a certain date, it would be obeyed only in so far as people could afford, by that date, to install sewage-treatment plant. It would not stop the production of sewage!

It may sometimes be possible to make people comply by

imposing exemplary penalties on those who continue to offend, but there is very real danger in proceeding that way. In the first place, the expense of taking offenders to court ought to be included in the cost of getting rid of pollution, and it is not easy to measure the suffering of a society sick with offenders, enforcement officers, parasite lawyers, and the indignities and inconveniences of penalties. The very existence of offenders probably means that the law is bad or unfair at least for some people.

It is usually impracticable to prohibit a monopoly, whether it be of cars in Los Angeles, or of an electricity supplier whose power station emits obnoxious dust and gases through too short a chimney. Legal wrangling that follows a prohibition can make it possible for the offence to continue for years if the way of life of society requires it because there is bound to be doubt about the outcome of the legal proceedings.

In considering prohibitions, and prosecution of offenders, we must not be confused by occasional deliberate surreptitious crimes like dumping of noxious waste in a river. The offender is probably attempting to avoid the expense of the accepted method of disposal and such obvious anti-social behaviour must be treated as a crime. But what if an industry is equipped to dispose of wastes in a particular way and its emissions become damaging after a few years because other people's emissions have been added to it? That a right to pollute had been established by past practice will certainly be argued if alternative methods of disposal are very costly. A public subsidy may be needed. If a factory exists, can its owner be blamed for disaster potential created around him by other people? It may be better for society to share the expense of new equipment with him than put him out of business or in prison, or both.

Sometimes attempts have been made to halt industrial activity when the weather causes pollution to accumulate in the air to intolerable concentrations. But it is usually futile. Some industries need a few days to bring them to a standstill, by which time the weather could have changed again. The present state of the art of weather forecasting does not justify compulsory closures of that kind (see p. 48).

In a sense all control of pollution is prohibitive, but in this section we are considering the use of prohibition as the direct and only means of bringing about change. Some prohibitions are quite easy to institute: for example, oil fuel from different parts of the world may have different proportions of sulphur. Venezuelan oil contains 4 per cent, Libyan only 1·5 per cent. The sulphur can be reduced by a rather expensive process and from Venezuelan oil fuel of lower sulphur content can be produced. It is therefore possible for a state to specify that no oil may be used which contains more than, say, 1 per cent of sulphur, and the regulation can be obeyed although perhaps at some extra expense. Such a law has been introduced in some states of the USA, notably New York, New Jersey, and Pennsylvania.

This law is easy to pass and is uncomplicated to understand. It is therefore attractive to legislators wishing to achieve some abatement of pollution. The objection to it is that it is undiscriminating and therefore uneconomical. It is known that sulphur dioxide (SO_2) is the main product of the sulphur in oil and that it is roughly reduced in proportion to any reduction in sulphur in the fuel. Since there are areas where the amount of SO_2 in the air is objectionably high, it seems logical to achieve a reduction by reducing the sulphur content of the fuel. This simple-minded idea is not incorrect, but it takes no account of the fact that many of the sources of SO_2 contribute negligibly to the SO_2 in the air in places where it is objectionable and contribute nothing significant to the damage. Thus, for example, the high concentrations of SO_2 in Manhattan are due to the use of high sulphur-content fuels in the houses of Manhattan itself. It has been found that a power station situated inside Manhattan contributes quite negligibly to the SO_2 in the streets there because of the tall free-standing chimneys from which it is emitted. Almost all the SO_2 in a street, which is where it matters, comes from the buildings in that street and in other streets very close by (within a quarter of a mile or less).

Almost the full effect of this law could have been obtained by prohibiting the use of high sulphur-content fuels in cities in which buildings do not have high free-standing chimneys. This would have left the consumers of about three-quarters of the high sulphur-content fuel unaffected by the law. Unfortunately, this slightly sophisticated idea is too complex for most people to understand because they have no guidance from the press. Indeed the press may have a vested interest in the matter because it is always easier to campaign for a simple idea than for a slightly more complicated one. Some politicians and journalists are more interested in the passage of some (token) legislation than in arguing about what is the best legislation.

One of the consequences of this particular law is that Appalachian coal, the sale of which had very much declined because of competition from oil, cannot now be sold at all in these areas, because its sulphur content is too high. This undoubtedly suits the salesman of low sulphur-content oil and natural gas, but it has worsened an already very serious economic depression in Appalachia, where unemployment has risen as high as 60 per cent in some areas. Thus the social consequences for West Virginia are out of all proportion to the benefits in the other three prosperous states, and even if the abatement of SO_2 in their cities is agreed to be necessary it could have been achieved at much less cost and with far less serious consequences for the coalfields. Furthermore, the low sulphur-content fuel is now being burned in places where high sulphur-content fuel could be used without detriment, and this makes low sulphur-content fuel less available for use where it would produce definite benefit.

Many people find it difficult to believe that a power station consuming as much fuel as a square mile of built-up urban area surrounding it nevertheless contributes only 1 per cent of the SO_2 at street level. They argue that if it is the biggest source of SO_2 in the area the most effective tactic must be to reduce its SO_2 output first. They fail to observe that by having very tall chimneys the effect of the power station's emissions has already been reduced to only a small fraction of what it would be if the chimneys were short. There is an emotional blockage which is compounded by the choice of language: big tall chimneys are always 'belching forth smoke and sulphurous fumes', small chimneys seem by their very nature to be small in importance, even if the building beneath them is large.

Air quality standards: a crime?

It has become fashionable to establish ambient air quality

standards, which are declared levels of pollution above which it is intended that the actual pollution will never rise. It has been assumed by legislators that this is an obvious and necessary first step in any law controlling pollution, and many states and countries have stated their standards. The situation is, in actuality, quite absurd for a series of ten quite separate reasons, any one of which is enough to spoil the idea of air quality standards.

1. The standard has to be established on some basis. No scientist can give a simple statement of what it should be because different organisms react differently to pollution. Thus, some forms of animal and vegetable life are killed by a high concentration for a short time, while other forms of damage happen only when a certain concentration is present over a long period, and much larger concentrations for short periods have no effect in those cases. The effect of a pollutant like SO_2 is very dependent upon the simultaneous presence of other pollutants. Thus, when smoke is present, the synergistic effect is such that the SO_2 may do five or ten times as much damage, or do the same damage at one-fifth or even one-tenth of the concentration as when the smoke is not present. Standards ought, therefore, to be expressed in a very complicated form specifying the levels permitted according to their duration and the concentrations of other pollutants also present. Such a specification would be too complicated to be useful even if it could be made with confidence, which it cannot.

To get an idea of the difficulty of establishing a standard on a sensible basis ask yourself the question whether the standard should be such that your grandmother ought to be able to go anywhere at any time without fear of being (a) made to cough, or (b) made to go to bed, or (c) killed by inhaling the air. Should the standard be for a grandmother who at the time of inhaling is (a) in normal health, or (b) in fine fettle, or (c) at death's door with bronchitis?

Such standards as are set are usually arrived at by people who are not expert in the subject asking people who are. Unfortunately, life for the scientist is made easier if he gives a simple definite answer to such a request, even if it is a very rough guess, than if he explains why no reliable answer can be given, and so scientists do sometimes give simple answers.

They thereby dispose of the problem as far as they are concerned and save themselves a great deal of trouble at the time, though not later on, as we shall see.

2. If one local authority, state, or country adopts a standard, on however flimsy a basis, it is almost always assumed by everyone else that the matter was gone into in great detail by scientists expert in the matter. Very often the scientists in question give their advice at second hand: thus a medico is asked for a standard based on considerations of human physiology; but he knows that man can withstand much more pollution than many other animals or vegetation, and so his reply is a guess he has obtained from a botanist or a vet. At last, after much guessing and argument at cross purposes, a compromise figure is arrived at (or perhaps the first guess is accepted because no one knows enough to contradict it), and this is announced as the standard. The authority will then presumably take steps to see that it is not contravened.

Other local authorities, states, or countries now feel impelled also to establish their standards and, as is customary in such cases, officials are asked to find what standards are accepted elsewhere. Having no basis for making a better estimate, they bow to greater expertise and follow the precedent. Soon, yet other authorities, noting the unanimity, join the chorus and the guess becomes hallowed by its universality and is soon officially recommended by a committee of the United Nations, none of whose members is able to reveal the absence of any justification for the standard chosen.

3. At this stage in the game the participants will want to know whether the standard is being violated within their territory, and they will discover that they either do or do not have some measurements. If they do they will have to decide whether to be satisfied with them, whether to make more in order to establish a network of smaller mesh so that no pockets of high pollution go unobserved, and what to do if they are not satisfied with the pollution levels already measured.

The basis for decision about what to do about the standard when it is established is quite different from the basis for the standard itself. It may be decided not to set up any measuring instruments because that would require action if the standard were exceeded, and this may be a very sound political decision in a country where resources are being strained. They may wish to wait for others to set a precedent before they worry about such relatively unimportant aspects of the economy as air pollution. This has not stopped them establishing a standard for quite other reasons, and it must be appreciated that the people who determine the standard are not the same as the people who decide that a standard shall be determined, nor are they the same as those who decide what to do with the standard when they've got it. If air pollution, though undesirable, is not the most pressing of their problems, it is not necessarily a bad decision to make no measurements. For purposes of public relations alone the standard may have been established, otherwise it would have been open to any awkward interrogator to ask why no standard had been established – 'Are we so backward . . . ?'

4. Let us suppose, however, that a standard is established and that measurements are made. In the case of SO_2 the actual value measured depends very much on the location of the instrument taking samples of the air. Almost every surface absorbs SO_2: painted walls absorb it less than porous stone, wet ones more than dry ones, growing vegetation more than winter vegetation; but nevertheless all absorb it to some extent. No sampling device can be far from these surfaces unless it is on a high tower, in which case it is not measuring the concentration close to where it might do damage. There is a very important difficulty here: the taking of samples depends on the purpose for which they are to be used, and the siting of the instruments would not be the same for all purposes. Thus for the purpose of determining the exposure of the community we need samples within the community, near the ground, near people; but for the purpose of studying the diffusion in the atmosphere we need samples taken from tall free-standing towers, or captive balloons, or from aircraft in flight.

There is another problem of considerable importance: should the sampler record an average concentration over a year, a month, a week, a day, an hour, a minute or a second? Each of these time intervals has merit for different purposes, but it is simply not practicable to do all that is necessary for every possible future purpose. In the end the number and siting of instruments measuring pollution is determined by where it is easy to arrange for them to be put and regularly attended to, and by the consideration that even in this age of data-processing by computer too much information can be an embarrassment, and will certainly be a considerable expense.

No network of recording instruments that is ever likely to be set up over a wide area such as a large town or county can be adequate. One can never know what happened between the instruments nor, if they only give the average for a day, can one ever know what high concentrations may have occurred for much shorter periods.

5. Let us suppose that in a state rich enough to establish the ideal network of recording stations, the pollution is recorded as having exceeded the standard at some place and time. We now have to assign responsibility for this illegal event. In a community with good legal traditions, guilt in a matter of this kind is not a political question, and so it must be established beyond reasonable doubt. This is almost impossible in a typical city in which every instrument will be surrounded by a multitude of sources and sinks of pollution, great and small, near and distant. It will necessarily be a matter of guesswork who was responsible for the excess, and in most cases a multiple responsibility will be assigned.

6. Let us suppose that, contrary to all experience of the complexity of almost every meteorological situation, it could be determined who contributed what to the pollution recorded. It would nevertheless be very difficult to argue the case in a court of law with enough cogency to make a prosecution worth while. A competent barrister could, in the circumstances, make the standard itself look like the bureaucratic nonsense it really is, and the problem of determining guilt in circumstances in which several sources together make up a total which exceeds the standard, while none individually contributed more than a fraction, is insoluble in terms of justice. Each could argue that it is not incumbent upon him to find out on every occasion who else is polluting the same air. Suppose, for example, that all was well when a furnace was lighted up and then unexpectedly because of a wind change some air polluted by others near by up to the value of the standard drifted over so that the effluent from this one additional furnace caused the standard to be exceeded. Was it the last source that was

solely to blame, or should the others, knowing what was being put out further downwind, have restrained their output?

From the scientific point of view these questions are purposeless but when a case enters the law courts they become very relevant.

7. Even supposing that all the obstacles so far described were overcome, or at least circumvented, we must question the value of the exercise. Instruments have to be sited in particular places, and such knowledge as we have makes it clear that the amount of SO_2 in the air varies so much from place to place, even in one street, that whether an instrument records a pollution level in excess of the standard is very much a matter of chance (unless the standard is grossly exceeded over a wide area, which presents an impossible legal problem). If the recording instrument had been sited perhaps two hundred yards away it might have incriminated someone else, or it might not have recorded an excess if it had been near to a growing hedge or some other dense form of vegetation that is good at absorbing SO_2. Without a much denser network of recording stations than is practicable there is no way of knowing precisely how representative the records of a particular station are. Even if such a dense network were practicable it would be neither socially nor economically desirable, because the effort ought to be put into methods of abatement. A sparse network cannot keep a continuous watch on air quality, and any prosecutions based on it would be chancy and unfair.

Perhaps a more serious objection is that if the standard is exceeded a prosecution of offenders does not necessarily stop the offence. It certainly does not stop it at the time, and if a prosecution were successful the offender might still not alter his practices but continue as before, hoping that the adverse weather which failed once to carry away his effluents satisfactorily would not occur again. If a second offence occurred, what sort of penalties should be imposed? Once we engage in discussion of this question we are so far from our original task of prevention that the argument is of a quite different kind. We become involved in the whole issue of the effectiveness of punishment, which, as we shall see later, ought studiously to be avoided.

8. The argument in favour of establishing standards is

more soundly based on consideration of new industrial and urban development than on trying to abate air pollution already plaguing our cities. Thus, in granting permission for the building of houses and factories a planning authority should take account of the air pollution that will be produced. If a standard is set, the plan will be based on the idea that never, in the course of the growth of the new town, will the standard be exceeded.

The difficulty of putting this idea into practice is to decide what the effect will actually be when the buildings are complete and occupied. In order to do any calculation it is necessary to have a theory about what will happen to the pollution emitted from a particular type of chimney in an area with buildings, trees, and hills of particular shapes, in winds of various kinds. In fact no such theory exists except for very special situations such as an isolated chimney on a large flat plain in a uniform wind. Even if such a theory existed it would still be necessary to find out what winds and weather were likely to be experienced in an area, because different places experience different winds. Furthermore, the winds of one year are not the same as another, nor are the winds of one decade the same as another, and we have no means of predicting what will occur in the next year, decade, or century, except in a very rough way by saying that it will be rather like our past experience.

But even the past experience is recorded in only a few places, and these have usually been chosen for their good exposure on mountains or in open plains. Very little in the way of detailed recording of the wind and structure of the atmosphere has been done in valleys, which is where most of our pollution problems occur.

9. If some sort of calculation or agreed procedure were adopted for giving or withholding permission to build new houses or factories, once permission has been given for certain building to take place it would not be possible to prosecute successfully anyone who operated according to the permitted plans, even if that caused the standard to be transgressed. If it turned out that the standard was frequently exceeded what steps should be taken? In all probability, once permission is given for houses and factories with approved equipment to be built, any measurements would be a great embarrassment

if a standard existed. If it turned out that half the standard level of pollution was never exceeded should the operators be permitted to make more on the grounds that they are being put to unnecessary expense to maintain their existing level of pollution control? Should other developers be permitted to come into the area on the grounds that the density of occupation had been kept too low?

If the predictions of the planners turned out to be correct, how long should the developers be allowed to operate without improving their pollution control? It might be in their interest to ensure that competitors should permanently be excluded from coming into the area. It might be in someone else's interest to insist on improvements because of the advantages of allowing denser development. Improvement for its own sake is certainly desirable, but how is a change in the standard justified? Thus the idea that the established standard somehow represents an absolute, valid for all time, like the 'minimum daily requirement of vitamin X', becomes a tool in the tussle for advantage. It will be exploited in every possible way. It can be represented as a standard such that no one is entitled to complain about pollution if it is not exceeded, just as we are not entitled to complain about other people breathing. A community living in a district with pure air will then have no means of maintaining higher standards of air quality than others which have chosen to engage in industry or busy commerce.

How large an area must be considered by an authority granting planning permission? Clearly there will be cases where pollution crosses international frontiers: in such cases, must account be taken of pollution likely to come from the other side in determining how much shall be allowed to be added locally?

10. Evidently a standard becomes a licence to pollute. It becomes a basis of argument for claiming permission to create pollution up to that level. And it becomes very difficult to improve it, because the basis of it, in theory at any rate, is that it is absolutely tolerable. As soon as we envisage the possibility that the standard might be changed we undermine the criteria for determining it in the first place. In theory. For, of course, we have argued at the beginning of this discussion, that there is no scientific basis for establishing a standard.

6

It can never be more than a rather vague statement of what a particular society is at the present time prepared to put up with, or an ideal towards which they intend to work. The difference between these two is so large that merely to announce a single standard, as *the* standard, is to act ignorantly.

It is not really practicable for different places to announce significantly different standards, for no one will wish to admit to a substantially lower ideal. People will always act according to what is possible both from the physical and economic point of view. The levels of smoke pollution acceptable in Westminster and the Yorkshire coal-mining town of Barnsley are different because of the nature of life and work in the two places. In different countries the differences can justifiably be much greater: a town in an under-developed country introducing industries for the first time, and needing the profits in order to make the next advance towards greater affluence and economic freedom for its people, can deliberately spare the expense of tall chimneys and dust arrestors if only because, at a lower level of affluence, the actual disadvantages to its people are less. Just as economic advance creates disaster potential, it aggravates the effects of air pollution and a wealthy society is driven to preserve its amenities more than a poor one.

This last statement is often adduced as an argument in favour of letting events take a 'natural' course: excesses will produce their own remedies. The argument is not valid because in the past pollution has merely been pushed out of the way or people have moved away from it. Only when there is no more room left does the cleaning-up process begin in earnest. Always the result has been more affluence and more total pollution to be disposed of although often the earlier excesses are controlled. Now, however, the world is getting full, and only by pushing the pollution on to the poorer sectors of humanity or by preventing them from participating in the affluence altogether can the rich section of humanity continue to increase its turnover and its exhaustion of resources, unless technology can save us from making pollution.

Air quality standards are a bureaucratic crime against the people. They bear little resemblance to what is or can be done in the circumstances to control or abate pollution. The assumptions behind them are false, and their existence leads to dishonest and irrelevant arguments. They distract attention from the effective means of advance.

In spite of all this, air quality standards have been adopted in some countries and states of the USA, and it is claimed that they are useful. Let us look at how they are used: it is assumed that the pollution produced by a particular chimney in the surrounding area can be calculated, and on that basis it is possible to determine what total pollution will be produced by several chimneys in a district. By a manipulation of the formulas it is thereby possible to make an allocation of the right to emit pollution to various industries in a district. It is also possible to have different standards for expensive residential areas and industrial districts and thereby, on unarguable grounds, to exclude certain types of pollution source from residential areas. The restrictions are actually specified in terms of the pollution each user of the air is allowed to emit: that is to say each is told what he may emit. The objection to this is that there is no scientifically acceptable way of relating the emissions to what they will produce in the ambient air simply because it depends on the weather which is too complicated to represent in a mathematical formula. Mathematical formulas are nevertheless used, and it is only the technical incompetence of a court of law that prevents them being argued there.

In Britain the same result is achieved without any reference to specific air quality standards. As in places where the standards are used as just described, a factory operator is told what he will be allowed to emit, but this is justified on the basis of the wide experience of the Alkali Inspector's staff of what is possible in industry and what kind of pollution is objectionable. The Alkali Inspector's judgment is accepted, and his staff are a useful source of technological advice to those to whom he grants licences to operate. In a government by pure bureaucrats precise specification as to how they shall proceed is necessary, and they can be told to use the scientifically phoney mathematical method of calculating from the ambient air quality standard what emission quality shall be permitted.

There are very few forms of pollution for which an ambient air quality standard can be used because of the rapid absorption of the pollution on the vegetation and buildings. And smells, which are becoming ever more important in modern industry, cannot usually be measured quantitatively, and therefore cannot be controlled by reference to air quality standards.

The 'best practicable means': the 1956 Clean Air Act in Britain

Earnest anti-pollution workers complain that the arguments of the last section serve only to undermine the most effective tool which administrators and politicians can use to get restrictions imposed. In reality, when measurements are made in a badly polluted area no absolute standard is needed to show that other areas are less filthy. In fact it is not really necessary to make any instrumental measurements at all of most kinds of pollution because our bodies are already equipped with detectors of very high sensitivity which are necessary for our personal survival.

There are very poisonous substances produced by some industries which need special precautions because they do not affect our senses. Beryllium compounds produced in the course of making fluorescent lamps cause lung tumours, and very strict prevention of escape into the air has to be undertaken. Fluorine compounds emitted in the manufacture of aluminium and bricks have to be kept down to small quantities because they cause abnormal and harmful growth of the bones and teeth of animals eating vegetation on which the compounds are deposited. Car exhaust must not be breathed inside closed buildings because the carbon monoxide, which does no harm at low concentrations, can cause death in high concentrations, and has no detectable smell. For the same reason town gas made from coal has to have a strong odour added to it in order that it can be smelled, otherwise people would have no warning of an escape and the carbon monoxide in it might kill people in an enclosed place such as a kitchen if the gas flame were accidentally blown out.

Most of these poisonous gases are well controlled because their presence is known from the chemistry of the industrial process which emits them. Many early forms of pollution such as non-ferrous metal poisoning in foundries were discovered by their effect on the workers; and because of the terrible consequences new processes are now much more carefully watched and controlled. We have learned the hard way.

The callousness or ignorance of some employers is a common human failing and must be seen in the light of the continuous series of experiments conducted by civilised societies on themselves. The damage done by smoking is widely known: in Britain, cigarettes cause the loss of about a million years of life every year, and this is well known by people in authority. Yet little is done to control cigarette advertising. This is mentioned not to arouse indignation: that is the purpose of the next chapter. On the contrary it is mentioned to draw attention to the apparent inability of communities to do anything about disasters of epidemic proportions if they are normal. If the casualties of smoking diseases could all be made to occur in one week of the year only instead of occurring monotonously day after day, public reaction would be immediate.

The great London smog of 1952 shocked the people and the government into action. If it had been the great Dewsbury smog, the action prompted by it would probably have been restricted to Dewsbury. But Parliament was in the midst of it. Furthermore it was preceded by weather which had caused rather more influenza than usual and chronic bronchitis had been aggravated. Consequently there were many elderly people waiting for the occasion of their death. At the same time as the smog, the like of which had occurred before, it happened also to be colder than usual and so more babies and old people were killed by bronchitis and influenza. It lasted for five days: quite possibly if it had lasted for only three, less notice would have been taken of it because by lasting for five it had a very serious effect on communications and crime.

The most startling effect was the resolve it gave to London to ensure clean air for the future. In three other cities, Manchester, Bradford and Sheffield, the resolve had already been made and some action taken to improve matters. The smaller city of Edinburgh, so accustomed to living in smoke that it had acquired the name 'Auld Reekie', took almost no action in the fifteen years following the introduction of the 1956 Clean Air Act.

The philosophy behind this Act was to use all practicable means to reduce pollution. In particular it was known that black smoke was unnecessary and that with suitable apparatus and training of boiler operators dark smoke could be prevented. A criterion was invented: an inspector could deem an offence to have been committed if smoke darker than 'Ringlemann 2' was emitted for more than ten minutes in any hour.

In the administration of the act the simplemindedness of most industrialists and public health inspectors was a great asset. Many obvious flaws in the system were ignored and a genuine attempt to improve the quality of chimney emissions was set under way. The inspectors continually warned obvious offenders, and threatened them with compulsory shut-down. Courses for boiler operators were held all over the country, and old careless practices became less common as more and more people discovered the economic advantages of improved combustion efficiency. The crude art of the possible was pursued.

The criterion – whether smoke was darker than Ringlemann 2 – was hopelessly inaccurate and inconsistent but it was simple and obvious. The apparent darkness of smoke depends on the direction and intensity of the illumination both of the smoke and of the background, and on the colour of both. Steam coming from a drying plant can, in certain lights, appear darker than Ringlemann 2, even though it can be the brightest thing in view when the sun is shining! Furthermore, if added air is passed up a chimney, the dilution of the smoke at the chimney top can make it lighter than Ringlemann 2, even though the amount of smoke emitted is as great as ever.

The criterion worked to a large extent because the people making the decisions in industry had almost no experience of out of doors observations and could not argue with the public health inspectors who kept their amused doubts to themselves.

The success, which has been undoubted, has depended on the tireless work of the inspectors who used their eyes and arts of persuasion. They taught industry, and industry taught itself to find the best practicable means to effect a substantial improvement. Various instruments have been installed in flues of the larger industrial plants which make records and give warning of too much smoke or of incomplete combustion, and ensured that the best possible conditions in the furnace were being obtained. The measurement of pollution in the air outside was not part of the programme of control at all, and such measurements of smoke as have been made over the years in our towns have merely served to confirm and roughly to quantify the decrease in smoke which has been obvious to our eyes and noses.

The control of industrial smoke from existing industries was only part of the operation of the Act. All new industries have been licensed. In addition to smoke, which can be almost completely abolished, industries consuming fuel emit large amounts of SO_2. This emission cannot, by any practicable means, be prevented. Therefore, the local government authorities were empowered to insist that chimneys should be of adequate height to prevent the occurrence of concentrated pollution near the ground. The result has been that although the amount of fuel consumed and SO_2 emitted has been greatly increased the actual concentrations measured have not: on the contrary, they have decreased.

Most large industrial works are not licensed by the local authorities but by the Alkali Inspector* who, like the local authorities, specifies any requirements and standards that must be met by the builder of any new works. He also supervises the operation of existing works. His staff are in almost continual consultation with industry and are well informed about all the latest effective methods of reducing pollution emitted into the atmosphere. Their job is to insist on the 'best practicable means' being used to achieve this, and they are able to do this because what is practicable is defined to include the economic practicability in the context of the industry in question. This means that normally the requirements insisted upon are not beyond the means of the company. Sometimes the requirements imposed are so stringent as to prevent a works being built in the place intended: this happens when the district has a high damage potential either because it is important to maintain high air quality for social reasons and it is necessary to preserve the amenities (as in the central civic area of a city) or because of sensitive vegetation, or because of the high frequency of adverse weather such as when the

* The Alkali Inspector's name referred to the first job he was created to do (to control 'Alkali and other Works' in 1880). In 1971 the name was changed to Alkali and Clean Air Inspector.

site is in an enclosed valley. When a development is prevented in such an area it is usually possible to find an area of lower damage potential in which to put the works.

Industry is required to use the best practicable means. It is therefore compelled to install more up-to-date equipment from time to time under threat of closure. Notice of a few years of this requirement is always given, and the details are always decided after consultation between the company and the Inspector. For this reason he has been described as 'in league with industry', the implication being that his activities are a charade conducted to keep the public happy. This criticism is not in the least justified: the Inspector's efforts are directed to the continual improvement of air quality by the best practicable means and in theory, therefore, industry is kept up to the highest attainable standards. The penalty for failure to comply with the Inspector's requirements is immediate closure without appeal, because the licence is withdrawn. If the Inspector were a ruthless dictator this system would be intolerable, but he operates under the direction of a minister responsible to Parliament, and must obey his directions. The minister is obviously interested in keeping industry fully at work and so a compromise is arrived at under which collectively we do as well as we can afford to do.

The most interesting aspect of the Inspector's work is that he is not required to reveal how he has decided precisely what requirements he will impose in a particular case. He is compelled by obvious criteria of justice to be fair as between one company and another, but there can be no argument, for example, about the height of a chimney. It must be at least as high and have at least as great an efflux velocity as he requires. The company can reason with him, and in practice the relationship is such that discussion always takes place. The Inspector gives free advice on how to attain the required standards.

The chief reason why this system works well is that although the Inspector has the power to prosecute and enforce closure, it is scarcely ever used. Prosecution is used, for example, when a demolition contractor persistently burns cables and other materials whose incineration on the site is forbidden. But prosecution is not used to control the normal operation of industrial plant because it is not allowed to operate at all if the equipment is incapable of achieving the required standard, and a warning is usually enough to prevent repeated misuse to the extent of contravening the standard required. The standards are emission standards, and no standards are specified about the quality of the air in the general environment.

The public is often worried when penalties are small, and is encouraged to believe that improvements in air quality are under way when heavier penalties are imposed. But the alternative interpretation of the imposition of punitive fines or imprisonment is that the system, whatever it is, is working badly. Unless exemplary fines cause the offence not to be committed any more they are useless; if fines continue to be imposed it means that the offence continues; if they are made harsh from time to time to frighten people into compliance with the regulations, clearly the public is not in favour of the law. It is better to withdraw a licence to operate without any prosecution or alternative penalty than to allow an offence and then take punitive steps to prevent its repetition.

To stop houses from emitting smoke it has only been necessary to ensure that the householder has equipment with which he can do his heating and cooking without making smoke. The Act permits a local authority to declare an area to be a Smoke Control Area. The order is confirmed by the government. The Public Health Inspector's staff inspect every house in the area and advise each householder that by a certain date he must install approved appliances and use no others and that he may only use approved fuels. The sale of other fuels is prohibited in the area, and if any prosecutions take place they are more likely to be of a fuel merchant for selling unauthorised fuel than of a householder for using it. But prosecutions are not normal.

The householder is informed in detail of what costs of conversion he may claim. The purpose of the subsidy is to ensure that no one can claim he is unable to afford the conversion.

The most important aspect of the working of the Act is the great amount of work necessary to ensure that everyone who is required to change his practices is properly advised how to do it. The appliances for house heating must be available and so must the fuel, and smoke control orders are not confirmed by the government unless they are satisfied about this. Before the Act was introduced enlightened people in the Alkali Inspector's department, and in the public at large co-ordinated by the National Smoke Abatement Society (now the National Society for Clean Air), had held conferences, exhibitions, and discussions with manufacturers, so that there existed in society a body of people who knew what to do and who could advise the politicians what law to pass. The politicians had no policy at all beyond fine sentiments!

To sum up, therefore, the experience in this case shows that success is achieved by getting everyone's co-operation; by having an advisory service available free to anyone who is required to take action; by not requiring anyone to do what is not practicable or beyond his means; by ensuring that the means are available; by not licensing works in places where the necessary standards cannot be met; by keeping the older sections of industry up to the best standards by imposing requirements on any new plant that is installed; and by imposing the best practicable standards on any newly built industry.

Incidentally, this system gives encouragement to industry to participate in the development of better methods of avoiding air pollution.

There is a danger of self-congratulatory complacency in the agreeable success of the Act, so let us look at its limitations. First of all it has been largely by good fortune that it worked. Britain established the core of knowledgeable people mainly through being the first to experience serious air pollution as the result of industrialisation. Second, we live in a well-ventilated country where the air does not stagnate as often or for as long as it does in Bohemia, continental North America, or in deep Alpine valleys. As a consequence we have been able to emit more air pollution with safety. Third, the Act has not been universally successful in respect of domestic smoke because many local authorities as late as 1972 had still taken no action at all. Because of this, an Act was passed in 1968 empowering the government to institute smoke control and to charge the cost to the local authority. This power has not yet been used. One of the reasons for inaction is that the air of some towns is not so unpleasant long enough or often enough for the people to feel that something needs to be done. Lethargy in the matter arises because people have become used to smoke as a natural part of the weather.

Fourth, and perhaps most important of all, there are some problems, such as the abatement of car exhaust in big cities, which we do not know how to solve technically without an expense which seems, at present, too great for the advantage to be obtained. In a sense we were very lucky that SO_2 has turned out to be less harmful than many people once thought it was. In fairness, there were others who said that getting rid of the smoke would render the SO_2 much less harmful, and that the increased sunshine resulting from smoke abatement would cause a greater dispersion of SO_2 because the ground would get warmer; but the actual reason why the Act concentrated on reduction of smoke output, but sought to reduce SO_2 only by building taller chimneys, was that this was all that was seen to be practicable. It was the best practicable means, and practicable means are not equally effective in all places and with all forms of pollution.

The Clean Air Act was an enabling act, and is in great contrast with those which seek to produce results through prohibitions, prosecutions, and punishments. It enables the authorities to enforce the Act, and they enable the householder and industrialist to comply.

Pricing pollution

Much favoured has been the idea that the polluter should pay for the harm he does. To this there are several overriding objections. First, if the polluter is allowed to pay he can buy the right to pollute, and he, not society, will decide whether the pollution is stopped. Second, if he decides to pay, he will pass the cost on to his customers: we are not dealing with rich people who buy the right to engage in a form of luxurious indulgence and so, since society will pay anyway, it might as well subsidise directly the cost of abating the pollution. There is much merit in the argument that if it is going to subsidise the operation society should take control of the business in question at the same time, and if this were accepted by a majority we would be saying 'stop polluting or be nationalised'.

Third, the problem of how much he ought to pay cannot really be determined because most air pollution causes a general deterioration in the quality and amenities of life, and does not produce only one or two simple consequences which can be costed. If we try to cost pollution we can find no clear logical boundary beyond which the cost attributable to pollution is certainly negligible. It is very difficult to draw a sharp line between natural weathering of stone and corrosion due to pollution; between the pollution due to specific sources and the general activity in a city; between the cost of restoring damage, the inconvenience of having to arrange the restoration, and a true recompense for the damage suffered; for it might be very costly or even impossible and perhaps undesirable to make good all damage. The costs can include the additional costs of running a household through the necessity to purchase more soap or do more frequent decoration, but should one be paid for the extra time required to wash one's hands or choose new wallpaper? How can the cost of aggravation of minor illnesses such as bronchitis, influenza, and common colds be assessed? Could it not be argued that if one area is persistently more polluted than others, houses will cost less there, and the purchaser is already automatically recompensed by the operation of market forces when he moves into it?

These issues are not in the least trivial in the context of trying to cost the damage done by pollution. Since one of the purposes of pricing is to organise some tax system to make the polluter pay, why should we not equally enable people to buy clean air by paying for it, that is by paying the polluters to stop; indeed why should not the polluters claim the right to be paid for not polluting? Or should people be allowed to buy the right to pollute by choosing to compensate those who become polluted?

It is impossible to decide upon any rights and wrongs in the matter once the idea is accepted of payment for pollution, one way or the other. If we let the market forces take charge and determine the amounts to be paid we arrive at a quite unacceptable position.

In the early days of the Industrial Revolution it was thought that if you did not like pollution it was up to you to get out of its way. So great were the immediate advantages to the polluter that he always had the initiative. When coal first came to London the smoke was strenuously objected to, but the advantages of coal were too strong and nobody was able to stop the increased use of it. The wealthy simply moved to the suburbs, and of course burnt coal there too. Perhaps a more telling example is in the spoliation of land, for coal-mining companies were able to buy at very low prices land on which to dump the unwanted material that was brought to the surface. Their profits were great because the cost of coal was merely the price of getting it out of the ground. The Industrial Revolution caused an ever-increasing mechanisation of the life of an increasing population until now the land has become so valuable that we seek to clear the dumps away, and the cost of restoration is very great. It falls on the community, and was never paid as part of the price of the coal.

Such careless spoliation of the land meant that profits were greater, and so one may suppose that industrialisation proceeded at a greater rate than if the early industrialists had had to clear up the mess they made as they went along; now we have a backlog of mined land, dirty rivers, and habits of polluting the air which it is very costly to rectify. Although we have much greater resources now to clean up the mess, we have an organisation of commerce which encourages people to arrange for the mess to be made somewhere else rather than not make it at all. The payment for the damage done in obtaining profits is not built into the system, and so the system will not work in a full world in which there is nowhere else to go. The question is whether a change in taxing or pricing systems can put this right.

But that is not the sole issue: there is also the question whether we should expect to be able to achieve our goals by pricing pollution. Society needs revenue, and at present it is mostly obtained by income tax (IT). For purposes connected with regulating competition between the different countries in the European Economic Community, there is much in favour of value added tax (VAT). The main argument in the pollution and population context against VAT is that it taxes products whose value is created by human labour more than it taxes machine-produced products based on fuel, and is bound to decrease variety and craft in production of goods and encourage substitutes for human employment. There has been a suggestion for a pollution added tax (PAT), or a container added tax (CAT), which would reduce the magnitude of the unwanted material produced in manufacturing processes or delivered as packaging with a wanted product. PAT would take account of the pollution produced by the manufacture and disposal of an article. CAT would seek to reduce rubbish production after it was sold.

From the point of view of their effect on the economy most of these taxes would follow the fate of IT. IT is progressive: the proportion of tax paid is greater on greater incomes and so it is recognised as being equivalent to a certain reduction of income: consequently, every attempt to use it as a means of redistributing wealth more fairly is frustrated by taking it into account in deciding income scales. Every increase in tax on the rich is parried by an increase in income differentials. The only way to make it effective in this respect is to make the tax confiscatory above an agreed maximum and negligible below; then at least we could curb the flaunting of great personal wealth. The high rates of income tax on high incomes also has a distorting effect on the industrial hierarchy. It reduces their number because of the high cost of their tax payments. If there were no income tax we could afford twice as many top men; whether this would be a good thing or not does not alter the fact that any attempt to reduce differentials by taxation reduces the number of top posts. It would certainly improve the atmosphere in universities if there could be more professors.

If it is revenue we seek we should recognise the sources of wealth and tax them. The main source of wealth is fuel, and if a nine times tax were put on it (resources usage tax, RUT) so that all mineral fuel were priced at ten times its present pretax value, and this were used to provide the entire revenue, we would then be in a position to dispense the benefits according to a positive policy. We could subsidise desirable forms of public transport, for example.

This is not as far removed from reality as it might seem, because the price of a litre of petrol in Europe is about five times the fraction of the average day's pay that it is in the

USA. This almost certainly is an important reason why cars are more economical in fuel consumption in Europe, and why more attention is given to engine performance than in America. Because the fuel is imported the high tax rate on petrol is accepted as unobjectionable. An examination of the tax on fuels derived from oil shows that it is almost entirely advantageous.

We are very careless with cheap fuel, and the discoveries of oil resources during the last two decades will make fuel and oil products cheaper. The consequent pollution will be greatly increased unless deliberate measures are taken to conserve resources.

Habits of life become deeply ingrained. If from tomorrow we were to charge a high price for the collection of domestic refuse we might soon find many people living in a squalid accumulation of it. After its amount had grown to a point where the cost of removal by the refuse service was too great for the householder it would become like a crippling debt, with the added disadvantage of its continual obviousness. Nevertheless, somehow we need to reduce the production of domestic refuse, and to increase the cost of packaging to the consumer is the only way of making it less saleable in the supermarkets. If we do not raise the cost of disposal we must raise the price of the material disposed of.

One must question the basis of the attempts to control excesses of modern society by pricing. Suppose, for example, we were to try to control population by pricing and introduce a birth rate added tax (BRAT) we would quickly find the ignorant poor more quickly driven into bankruptcy than at present. The whole idea of pricing and taxation is either to price away the undesirable or to provide a source for public revenue. In the case of pollution, the first of these two is bound to fail because most wealth is made by operations which make pollution: therefore the people who it is intended to discourage are the very ones who can afford to carry on as before, and if we stop them we may stop the source of wealth for everyone.

We need therefore to price as high as possible the commodity whose waste and use cause direct pollution, and with the revenue pay for the prevention of the polluting by-products of industry.

An important effect of direct taxation of natural resources would be that as soon as a new resource became available, such as North Sea oil, there would be an automatic increase in revenue without any changes in taxation levels. The new revenue would be used directly to support public works such as clearing of derelict land or a pollution abatement programme. Under present arrangements it has been assumed that there would be increased revenue produced by greater industrial activity, mainly in the form of increased receipts from income tax. But there is no assumption built into that system that new wealth is a bonus that ought to be used for public purposes: on the contrary, there is always increased pressure for a reduction in tax levels, as if less should be taken of people's hard-won earnings. In fact most of it was earned by the using up of fuel resources: we steal it from posterity, and the thing stolen is non-returnable.

Most of the pollution from oil is due to its low price. We are unduly extravagant and careless with it: it is spilled on our roads, and many accidents by fire would be avoided if the commodity were more valuable. Old sump oil is thrown away: it could be cleaned and re-used, but the cost does not make it a worth while proposition at present – new oil is too cheap. Furthermore, in our society what is new is given more trust than something old but reconstituted: and so it is believed that cleaned oil would be difficult to sell.

Pricing the primitive sources of wealth such as minerals, fuel in particular, and extravagances like packaging seems to be the most promising way to reduce the pollution we make, and certainly much to be preferred to pricing the pollution as such.

The polluter must pay

This is a slogan which is widely thought of, particularly in the European Economic Community, as representing a basic principle in the handling of pollution problems. It is based on the idea that no one should gain commercial advantage by being able to make pollution which someone in commercial competition is not allowed to make. Thus if one country were to regard the abatement of pollution as a public responsibility and paid the cost of all dust abatement machinery at a cement factory, it would give it an unfair advantage over a cement company in another country where the cost of dust control had to be borne by the producer, and therefore passed on to the customer.

Equally, if one country were to say either 'we have no pollution laws', or 'we regard pollution control as a public expense', companies would be inclined to build new factories in that country rather than in one with stringent anti-pollution laws. To avoid this the EEC has accepted the so-called principle that 'the polluter must pay' as a statement that the cost must always be borne by the polluter in a similar manner in all countries.

This seemingly good principle is really too simple for a very complex situation. Pollution control may be much more important at one factory than at another because of a higher population density nearby or because the agriculture surrounding it is more sensitive or because the weather and lie of the land made dense pollution more frequent. The implication of this commercial principle is that every other factory using the same production methods ought to be compelled to employ the same costly methods of pollution control even if the cost far exceeded the damage done by the pollution, in order that it should have no commercial advantage.

To carry this purely commercial criterion to its logical extreme, the tomato growers of the Po valley ought to have to grow their produce under glass because that is what the Dutch tomato growers have to do, and they are thereby put at a commercial disadvantage. Carrying this one step further, assuming that the glasshouses are installed in the Po valley, the Dutch growers ought deliberately to smash their glass once every summer, or be compelled to use much more expensive thick glass than is necessary because the hailstorms of north Italy are much more severe than in Holland and glasshouses of Dutch design would not last long in Italy.

The commercial principles of the EEC are in many respects as bureaucratically ludicrous as that. They are based on the idea that if a product is more economically produced in one place than in another this ought to be because the production is more efficient or, if it is due to natural causes, then all the production of the product in question for the whole economic community ought to be located there as a result of price pressures which are thought of as natural.

The fallacy of this argument is obvious in the case of pollution because pollution cannot be commercially priced. The illogicalities that result from the logical application of the principle show that it is a silly principle, i.e. it does not work. This throws doubt on the whole concept of letting commercial criteria be decisive. In the environmental context it is impossible to balance the costs to us now with the costs of our actions to posterity or even to ourselves in two or three decades. The decisions which impinge on the environment must be political, and not purely financial. Our commercial criteria have been developed in an expanding economy, and do not relate to a situation where limits on continued expansion are imposed by resources and space.

New technology: our inability to take advantage of it

The main purpose of pricing pollution is to determine what causes the most costly damage and to compare the cost of abatement with the probable benefits which will follow. There is an assumption that conditions will remain more or less the same for a few decades, otherwise estimates for the future are likely to be unreliable. All the suppositions made in any calculation of costs are almost made meaningless by the invention of new technology and the discovery of new natural resources. We are never ready for new inventions and certainly have no proper plan for their use before they are discovered. Yet new resources and new technologies have more impact on our economy than anything else.

The reaction of 'Industrial Revolution Man' to new discoveries is to exploit them as quickly as possible. The new technology is quickly secured in the clutches of the conventional centres of power in business. It becomes committed immediately by adding to the tempo of trade before governments have been able to give any thought to the question whether new inventions or new resources should be handled differently.

Much of the argument against attempting to slow down economic growth to reduce pollution problems is based on the assertion that we cannot afford to rehabilitate slum areas, to clean up rivers, to re-landscape and cultivate land destroyed by mining operations and to clean our air, unless we have wealth to spare for those purposes as a result of economic growth. The backlog of the past, it is said, is so great that we must create a surplus of wealth or we shall have no room for manœuvre. Built into this assertion is the assumption that people cannot be made to give up anything in order to achieve a public good, and proposals for public works are often frustrated by the assumption that tax payers would have to give up something. The behaviour of people during a war shows that in fact they will give up a great deal very willingly and accept a much greater equality of wealth and a reduction in extravagant consumption of goods if the public need is clear. During peace they accept a high level of taxation for public purposes, and it is a fiction that in the twentieth century people will only vote for a political party on the basis of the material prosperity promised for them. Certainly there have been brief periods in which this has appeared to be true, but a close examination shows that during these periods the mainstream of propaganda has been an invitation to gluttony, acquisitiveness and fun through waste. There is no doubt that the propaganda was effective mainly because of its dominance: at times when contrary exhortations have been as loud, as constant, and as ubiquitous, their effect has been at least as great. It is not that people are naturally gluttonous, but that if they are never told of the objections to gluttony they will pursue it when it is easy and widely advocated.

In the past they were told it was wicked, but we know that the retribution which the preachers promised did not usually go in the right direction if, indeed, there was any.

But now people are becoming sick of it. They are ready to accept re-allocation of existing resources to preserve our environment and rehabilitate the damage of the past. Unless we make a start in putting aside some of our resources for this purpose we shall never have the organisation to do it with the future wealth which we assume will come from economic growth.

The argument that we need economic growth to put us in a position to clean up the ravages of the past and prevent further ravages is really nothing more than an excuse for not re-thinking our position. Those who advocate economic growth do so on utterly conventional lines which means that at the same time as areas of new prosperity appear there will be depressed areas, periods of acute unemployment, and patches of social distress occasioned by the changes in society. All this is in the industrialised countries; meanwhile, far greater un-employment, famine, social disharmonies and conflicts will afflict the densely populated areas of the less developed world.

The fact of the matter is that no plan whatever has been made by any government to use future and still hypothetical wealth in a new and different manner. The wonderful dis-covery of natural gas and oil under the North Sea is dis-appearing from the sight of the ordinary man; it has been assimilated quickly into the system; there is a temporary alleviation of some economic difficulties but the system remains unchanged and will merely produce greater contrasts of wealth and place greater power in the hands of those whose methods have produced all the problems of the past.

Many of our problems are produced by new technologies. Many investments of the past turn out to have been too great. The railways have been described as 'overcapitalised', which means that they were so well planned and built that they were capable of providing a service for much longer than the demand continued. Their structure became obsolete in the age of road and air transport. By contrast, the sewers and water mains built nearly a century ago have proved to be so well constructed that we have almost got out of the habit of thinking of their renewal. When they are renewed they will be re-constructed on the assumption that they are to last for as long as possible, probably another century. This is a case in which it is rather easy to justify building for such a long term, but the pace of technological innovation is such that in most cases we plan for only a generation or less ahead. So great is the return for investment nowadays (i.e. interest rates are so high) that a profit on capital investment is usually realised within a dozen or so years. A positive and deliberate decision has to be made to include unconventional uses of wealth in a plan extending beyond a generation hence. So uncertain and rapid is the turn of political and technological development that instead of planning to use our future wealth to build a different society all our governmental effort is consumed in dealing with new daily exigencies and the style of our economics is not altered.

It will be obvious that although we may not exploit the possibilities of the new technology as we would wish, nevertheless a very considerable use is made of it. From the conventional viewpoint of 'Industrial Revolution Man' we certainly take very great advantage of new resources and inventions. We fail in the two contexts of justice as between men who are not immediate neighbours and of planning for the future in which there will be no more room for continued development in the old style, i.e. justice between us and our successors. Plans for the use of new wealth when it comes ought to be quite specific, and we should not wait until it is available before making those plans.

The mood of revolution

So entrenched are the habits of thought and channels of commerce that the revolutionary mood which becomes ever more widespread in the young is regarded by many responsible adults as a kind of intellectual luxury which offers nothing for man's future. The words 'exploitation', 'fascism', and 'colonialism' refer no longer to old-fashioned situations of the inter-war years when their meaning was fairly obvious; they are now used to describe a new appreciation of the power structure of society. It is important to appreciate that attitudes which seem extremely revolutionary are becoming more popular simply because they are becoming more widely understood in an ever more literate world. It is easier from a revolutionary and iconoclastic viewpoint, than from a conventional one, to understand how such terrible things as slavery and excessive pollution were ever acceptable to those who deliberately practised them.

Ideas take hold, and young people of today in our wealthy society are far freer to pursue the new analysis to its proper conclusion than in the days when they were forced to conform in order to establish any sort of independence in living. The very acceptance of the idea that no citizen shall be allowed to fall into poverty has made the poor see themselves in a new light: they are citizens who have hitherto been deprived of their rights. By whom have they been deprived? The deprivation is economic because the control of capital is in very few hands and is dispensed according to very narrow criteria which do not include the public interest explicitly. There is no democracy in the allocation of economic power, and that power determines where most of the energy of society is directed.

At the root of the revolutionary mood is a rejection of the idea that economic advance is an adequate satisfaction for every individual. It is not that anyone rejects economic advance: everyone regards it as desirable, but the individual does not wish to receive it as his full entitlement as a member of society which every year has more to hand out: he wishes to choose and mould his own environment even if it means less material wealth, and perhaps even poverty. There is a widespread feeling that to join in the rapid material advancement as a willing participant is to sell one's soul, to kill one's identity, to consume what someone else chooses to offer in plenty and await the next round of beneficence at the hands of the captains of industry whose stranglehold on the government increases daily. The kind of fortune we have agreed to seek depends more and more on industrial man's increasingly unscrupulous exploitation of the natural world, and at a certain level such exploitation becomes as revolting as exploitation of man by man.

Having long since agreed to participate in acquisitiveness, people are now beginning to read the small print which says 'we, the captains of industry, arrogate to ourselves the right to tell the people what they are demanding'.

The mood of revolution is not so much against poverty and slavery as it was in the past. Now it is against the arrogant assumptions of those who choose to rule: against the right to dismiss an employee, against the entrenchment of husbands' rights over wives, of parents over children, of rich over poor. Basically, it is against the right to decide what someone else shall be satisfied with. Against the creation of demands, against the announcement of public wishes discovered by opinion polls based on loaded questions. Against IQs, against being trained to serve as someone else's employee, against being told to fight someone else's war, against documentation, categorisation, and computerisation of the individual, against rejection, against punishment.

The revolution is to give to everyone those freedoms which the Industrial Revolution has given to some, and for dignity for small and unimportant people when they make an intellectual without an economic advance.

It appears that power in the economic system blinds people to the value of anything but economic advancement, so that those in power cannot understand why everyone else does not wish to join in their plans for more material possessions for all, or at least for some, nor why people are not satisfied to be told that in such and such a way lies a higher standard of living.

The revolutionary mood is not new – it has always been part of human life – what is new is the scale of the enlightenment of those who seek freedom. Many communities in the past easily reached a state of outraged indignation at their poverty and powerlessness in their own lives, but only since Marx has there been a worldwide attempt to understand how power corrupts and makes people callous, arrogant, and cruel. In particular, the revolutionary mood makes people more aware of who their allies are. The Black Power movement in the United States has passed on from the campaign for freedom from slavery and for civil rights into a demand for more control by black people over their own affairs. It has proved to be quite inadequate to be integrated as equals into the system because the system itself is founded on racialist assumptions: it is the same system which supported discrimination, it is dominated by the white man's adulation of his own achievements hiding what he did to black men and counting for nothing the valuable attributes of other civilisations. White man has to be made aware by black men how arrogant he has been for six hundred years. White man does not yet understand that to bring Christianity and technology to others brings no entitlement to rule and proves no superiority of individual merit.

White man has much to learn about the structure of his own civilisation from the new articulate black men who are able to see history in terms of the prospects given to present-day man by the material wealth of white men. The spokesmen of the South African government are quite explicit: 'We have given them [the Zulus] Christianity and Western Technology and they should be grateful; they are better off than Africans in states with Black governments.' These spokesmen believe that God has granted them their position of dominance and say so quite explicitly. But the white business man does not hold such fundamental beliefs. He is convinced that because he has been successful within the law he is entitled to the fruits of his labours. He recognises no debt to others because, as he sees it, all co-operation on their part has been as voluntary as his own; the idea of the legal commercial contract is to establish a pretence of equality of the contracting parties and of freedom in bargaining. They are unable to admit, and probably cannot see, the gross imbalance of power which determines most contracts, nor can they understand that the law (and the civil and military power behind it) does not operate equally on behalf of all observers of the law. They are unaware that the adulation of wealth compels all the organs of society to favour the wealthy. Sir Thomas More saw this clearly in his *Utopia*:

> Whosoever be infamous for anything is made to wear golden earrings, gold rings upon their fingers, and chains of gold around their necks, and their heads are tied around with gold. By all means possible they make gold and silver a sign of reproach and infamy.
>
> ... they marvel that anyone is so mad as to count himself nobler because of the finer thread of the wool he wears, and that gold, of its own nature so unprofitable a thing, is in such high esteem: indeed that man himself by whom and for whom gold is put to use is held in much less esteem than gold. A lumpish blockhead churl, with no more wit than an ass, shall nevertheless have many a wise and good man in bondage because he has a heap of gold. Which, if it should be taken from him by misfortune or by some subtle wyle of the law (which no less than fortune raises the low and plucks down the haughty) and given to the vilest slave and utmost drivel of his household, then very soon he shall go into the service of his servant as a mere bonus added to the money.
>
> But they marvel much more at, and detest the madness of those who, although not in debt or danger from rich men, do give them almost divine honours for no other reason than

that they are rich, knowing all the time that as long as they live they'll get not a farthing of their heap of gold.

These and suchlike opinions they have conceived partly by education, being brought up in a commonwealth with laws and customs utterly different from such folly.

The advent of health and adequate food is enabling more and more people, not captured by the acquisitive philosophy, to perceive the extent of its power over compassion and reason. Even over sex.

Strength comes from knowledge, knowing who you are, where you want to go, what you want, knowing and accepting that you are alone on this spinning, tumbling world. No one can crawl into your mind and help you out. I'm your brother, and I'm with you, come what may, and against anything or anybody in the universe that is against you. You'll meet women and they will say they are with you, but you'll still be alone, with your pain, discomfort, illness, elation, courage, pride, death. You don't want anyone to crawl into your head with you, do you? If there were a god or anyone else reading some of my thoughts I would be uncomfortable in the extreme.

Strength is being able to control yourself and your total environment – yourself first, however. (28 December 1969)

There are millions of blacks of my father's generation now living. They are all products of a totally depressed environment. All of the males have lived all of their lives in a terrible quandary; none was able to grasp that a morbid economic deprivation, an outrageous and enormous abrasion, formed the basis of their character. (4 April 1970)

The blanket indictment of the white race has done nothing but perplex us, inhibit us. The theory that all whites are the immediate enemy and all blacks our brothers (making them loyal) is silly and indicative of a lazy mind (to be generous, since it could be a fascist plot). It doesn't explain the black pig; there were six of them on the Hampton-Clark kill. (21 May 1970)

George Jackson, *Soledad Brother: Letters from Prison* (Cape and Penguin, 1971)

Gandhi said that man should free himself from the persuasions of sensual gratification. Many monastic orders say the same. George Jackson, in prison for many years, deprived of such gratification, learned to master his resentment and understand how people become constrained into a viewpoint convenient to the power structure of society. 'The power structure' is a phrase that we cannot really escape because it is not so much the deliberate and concerted wish of those with power that forces society along certain paths but the fact that the system enrols their support without them appreciating what is happening to them, still less with them appreciating what the system makes them do to others. The rebels and heretics among them are made to fear that they will be cast aside and made ineffectual.

We have already pointed out that a good case can be made that the capitalist system will produce remedies for its abuses, and will cure pollution problems of its own creation; but even its leaders are powerless to prevent some of the problems from arising. We may see this in the problem of the disposal of dangerous chemical wastes: a manufacturing firm has by-products, such as cyanides, which it wishes to get rid of; to do so quite safely is very expensive, and so it resorts to illegal and surreptitious dumping, perhaps at night on remote refuse tips or even on moorland, and the danger only occurs years later when the containers are penetrated by corrosion or damaged by digging or bulldozing operations carried out without knowledge of their presence. Vegetation and animals may be killed, and even urban water supplies made dangerous when the poison escapes. The remedy sought by much of society is to prohibit the dumping, to raise the penalties imposed upon offenders, and thereby create a crime which the police will have to spend more of their limited resources in detecting. This remedy is a very obvious one, but it simply raises the stakes. The manufacturer may conspire with some employees to continue to break the law undetected. He is, by the system, encouraged by the financial gain to himself, made to consider this possibility especially if he has reason to believe that his competitors are subject to the same pressures. The fact that such dumping takes place, together with many other examples of the deliberate release of noxious substances into rivers and public drains, proves that people are under pressure to do it. The pressure is financial reward, and it works very effectively to reduce the consideration of the public interest in the minds of managers. This pressure operates on all advertisers, and standards in the advertising industry continually rise and fall as public concern rises and wanes following abuses and their correction.

In making laws to prevent these abuses we tend to think of methods of correcting the attitude of the perpetrators by threats of reprisal but we do not think, usually, of removing the incentive to crime inherent in economic and social structure. By hallowing the system, society consolidates the incentive, and in the full world we are now approaching it is necessary for us to remove the incentives rather than increase the role of crime and its detection, and the problems raised by what to do with the criminal, in our system.

In order to improve public health we provide free lavatories and a free refuse disposal service. Should we provide a free public service for the disposal of dangerous waste chemicals? This is an orthodox approach, and it would solve the problem of the pressures which make people act criminally. It does not, however, provide any incentive to solve the technological problem of operating the process without producing the dangerous by-product. We need to go one stage further and create pressures which will make us advance our technology on the assumption that every aspect of the manufacturing process and its effect on our world must be considered. The cost of disposal must be transferred to the consumer of the product, which is probably most of us, or the whole manufacturing operation must be made a public responsibility. The revolutionary view is that the industry should be nationalised or subject to strict public control, and the right of individuals to make as much profit as they can out of a public need should not be recognised as basic but merely conceded in special cases for reasons of public interest.

It is important that this kind of revolutionary viewpoint should be more widely discussed because, as long as the profit motive dominates most of the world's commerce, it will be difficult for one nation to create new pressures and incentives without cutting itself off from the rest, and to do that, or to threaten to do it, requires a strong political unity and will throughout the whole community which will be very difficult to obtain.

Radioactive pollution is very dangerous: these arguments suggest that industries using nuclear energy and creating radioactive wastes should not be in private hands – not under the present system of incentives at any rate.

The manufacture of cement is an example of the inertia resisting change. The air pollution produced by cement works is tolerated because it would be commercially very dis-advantageous to replace existing works with new ones of a different design which are not well tested on a large scale, while the existing ones have many years of useful life in them. Who is to judge the magnitude of financial advantage when there is no valid costing of the effects of pollution or, to be more correct, no financial measure of the advantages of not having it? We can never cost improved amenities because the amount we are prepared to pay for them depends entirely on how rich we are. Who is to promote research and development into methods of cement manufacture which do not pollute the air? If it is left to the manufacturers or their associations to do this, how do they, under the usual commercial pressures, weigh the value of the public interest in clean air as against cheap cement. If the industry is taken into public control how do we ensure that value for money is not disregarded in policy making?

The revolutionary viewpoint in anti-pollution legislation must strike much deeper than the well-understood argument between capitalism and socialism. That argument arose from the conditions of the Industrial Revolution and the desire of all men for material advancement out of poverty and to share the benefits of science and technology. Any criterion for handling the problem now must be based on the assumption that in two senses our society must become stabilised.

Expansion of population and increase in the tonnage of our consumption must cease soon, and we must envisage the criteria that will operate in the future if we are to work to-wards them intelligently and deliberately, and not wait for an accelerated succession of catastrophes to force new attitudes upon us.

If our stabilised society is to possess the exciting vigour of the age of industrial expansion we need to visualise the challenges that will attract men of independence and initiative. This is a very considerable task. If the financial advantages of exploitation and crime are to be eliminated by a more egalitarian view of incomes and the place of capital and property in the system, it is no use expecting men to be content with dreamy exercises in the arts, with sensual gratification of any kind, for the initiative will be grabbed by those who find satisfaction in hardship and personal risk. There is plenty of scope for these qualities in the period of revolution which will destroy the dominance of financial wealth in our evolution, but then what? Perhaps we shall not be able to see the next stage so long as we have intolerable racial and economic wickedness practised in the name of legality.

Capitalism cannot solve our problems because it eliminates idealism: it throws young people into a world with a power structure under which survival only seems possible by submission or corrupt participation.

Until a new idealism operates, every new gain in wealth such as the discoveries of oil in Libya, Nigeria, Alaska, and the North Sea will continue to be absorbed into the system and not used for essentially new purposes – gifts from nature though they be.

Scaring the public: lead, carbon monoxide, cigarettes, mercury, DDT, SSTs, 'Wolf!', 'Fire!', 'Help!'

Perhaps the most perplexing aspect of the problems of a technological age is that the public can be easily misinformed. Every danger depends so much on geographical, meteorological, and social circumstances that it can be played down or easily exaggerated. If pregnant women had never used thalidomide it would probably still be in wide use; if the weather of a few great cities like Los Angeles had been different, the car problem uppermost in people's minds would be traffic congestion instead of air pollution; and while science can cause overpopulation by advances in medicine, it can invent the Pill and two extremes become possibilities.

Prophets of doom are likely to be proved wrong in proportion to the extent to which they are taken seriously enough for society to be jolted into action, but that is only a long-term aspect of the problem. We have short-term scares, too, and we shall now discuss some of them which have done a disservice to the cause of sensible pollution laws.

It has long been known that painters and plumbers are liable to suffer from lead poisoning as a result of handling lead-based paint and lead pipes. The lead may enter through the skin, through the mouth via the hands, through breathing when old paint is being burnt off or when pipes are soldered. Children have been known to suffer lead poisoning because they chewed the paint off the window sill!

Every one of us takes in and excretes some lead. All lead to which we are exposed is not absorbed into our bodies, and not all lead which passes into our bodies does harm. It has certainly not yet been established that the lead in car exhaust does any harm to human health, and there may be some very odd reasons for this: for example, lead has been detected attached to carbon and other larger particles in exhaust but not freely on its own at particle sizes which can penetrate into the depths of the lungs. Certainly lead is not a component of car exhaust which causes us any discomfort or nausea.

Lead and other metals produce poisoning in very odd ways. Painters overcome the problem by drinking plenty of milk; this has the effect of driving the lead into the bones, where it does no harm. If a painter changes his job and gives up the extra milk, the mechanisms which carry the lead to the bones cease to operate and the accumulated lead comes out of the bones into the blood stream and produces the symptoms of lead poisoning. Thus the painter 'gets lead poisoning' when he gives up the job that causes it.

In brass foundries there is a disease from which the workers suffer which produces a shaking of the limbs. After a short exposure to the metal dust and vapour a worker becomes immune to further doses so that most of the time he does not suffer at all. The worst time is on Monday afternoon: it might be thought that after having Saturday and Sunday off work he would have returned to work rested and restored to full fitness; but the immunity has gone and the fresh exposure on Monday brings on the ague again. When a six-day week was worked, the exposure on Saturdays caused the immunity to be retained until Monday, and so for the workers' comfort it would be better to take a day off in mid-week instead of Saturday. The five-day working week made Monday a misery.

Lead scares exist because lead is known to be poisonous in sufficient doses, and we are warned to keep our children from eating it. The assumption that all lead is bad is not necessarily correct. The fact that there is some lead everywhere in the air in towns is not necessarily a threat to health. Many people see in such statements a dangerous risk that the public may become poisoned on a grand scale over a long period – 'because it is cumulative'.

Most poisons are cumulative in the sense that they achieve a concentration in the body in equilibrium with the intake, and some achieve dangerous levels when the intake is fairly low. Equally immunity can be built up: there is the well-known method of murdering an acquaintance – you build up a resistance to arsenic by a gradual increase in your own intake until you can withstand a dose which is lethal to ordinary people; you then share a drink with your victim.

A good example of a very poisonous gas that is a cumulative poison is carbon monoxide (CO). It is much more soluble in blood than oxygen, and when it combines with the haemoglobin it prevents the blood from carrying oxygen to the brain and thereby causes death. If you breathe air containing a constant proportion of CO the amount in the blood rises to an equilibrium amount in a period of several minutes. If you then begin to breathe air containing no CO the amount in your blood falls to negligible amounts in an hour or two. You make a complete recovery from mild CO poisoning simply by breathing it out again. The body cannot sense its presence in any way and it is quite harmless, in small quantities, to people in normal health, and the amount that can be taken in by breathing the air of a traffic-laden street is less than one-third of what is absorbed by a cigarette smoker. The effect, even on the smoker, is to reduce the oxygen carried by a given quantity of blood by the same amount as going up a mountain about 1,000 metres high. There are many cities in the world – Denver, Ankara, Johannesburg, Mexico City, Darjeeling, etc. – where people live far above that height. Indeed, a little oxygen starvation may be a cause of a feeling of elixir, and it may be the slightly sedative or tranquillising effect of CO that is attractive to some people about smoking.

CO is dangerous in large quantities and it is easy to cause death by shutting a person in a garage with a car engine running. Even small quantities can be dangerous to people who have no reserve in their system and whose lungs are so damaged by emphysema (structural damage due to coughing and blockage by mucous) that they can only transfer small quantities of oxygen to the blood at the best of times. Or their heart may be unable to pump the extra blood needed when its oxygen-carrying capacity is reduced or, if the weather is very cold, they may be compelled to breathe in and out so much cold air (which is mostly nitrogen that simply goes in and out again) that their body temperature is reduced. A person who has to sit and gasp for breath every few steps on going upstairs can be killed by a dose of CO that a healthy person would be quite unaware of.

Much of the early factual literature about air pollution stated the tonnage of CO emitted into the air, as if this poisonous substance were a public danger. In fact it has never been a public danger except in obviously enclosed places such as garages or where a stove has a flue which leaks into a closed room or, of course, the kitchen gas oven.

A device which is now used to improve the quality of car exhaust is designed to burn up any fuel which comes out of the cylinders unburnt. The purpose is mainly to burn up the unburnt hydrocarbons because they are a major contributor to Los Angeles-type smog, and convert them into water vapour and carbon dioxide. It so happens that the easiest way to measure its effectiveness is by the extent to which it reduces the amount of CO, which is converted into CO_2. Journalists and TV teams visiting the testing department almost always come away with the simple idea that the main purpose is to decrease the output of poisonous CO. It should be remarked that even where Los Angeles-type smog is not a problem these afterburners do make the exhaust less unpleasant to breathe because it is the compounds of hydrogen and carbon (and a little sulphur in some cases) which are smelly and nauseating.

As far as carbon monoxide is concerned smoking is very much worse, and to have a smoker in the car is as bad as having other traffic outside it. Even to work all day in the old Mersey road tunnel from Liverpool to Birkenhead raises the CO level in the blood by a quite harmless amount.

Public health is damaged many times more by cigarette smoking than by any form of air pollution, and if we are genuinely worried about public health it should claim our first attention. One million years of life are lost in Britain every year by smoking: about 100,000 people die on average about 10 years early because of one of the smoking diseases – cancer of the lung, the stomach, and the bladder; chronic bronchitis and attendant influenza and heart failure; ulcers and heart diseases. Furthermore, the tax on cigarettes scarcely pays the cost of treating these sufferers, and certainly does not pay for their lost work. A cigarette shortens a person's life by about the time taken to smoke it, and in that time the valuable work they could have done is far more than the total cost of the cigarette.

The public is not scared of cigarettes because of familiarity with them: it can be scared about CO in harmless amounts – in amounts less than one-third of what smokers give to themselves. If we could accumulate all the smoking deaths into one or two days a year the dramatic effect might shock the public into more strenuous action to stop young people smoking. For example, if the names of all who died from smoking diseases were collected and published on 1 January as a casualty list the public would be horrified. Beside them all deaths on the road, from suicide, and from fire would look few (about one-tenth). And deaths by murder would look utterly trivial (about one-five-hundredth). Parliament has spent many days in debating the death penalty and murder statistics which refer to one or two hundred people at most not because of the fear of being murdered but because of the moral problems. So far the cigarette manufacturers have managed to prevent smoking from being seen as a moral problem by the public: they merely live by exploiting to the full a human weakness, they seek to create addiction in the young and wound and kill the weak in so doing.

General public ignorance and failure to understand an issue is sometimes equalled by official ignorance. We cannot eliminate all traces of all substances, which in excess would do us harm, from our breath and our food. Indeed some such substances are

8

necessary. Mercury has never been a cause for public concern on account of its use in dental fillings. Provided it is handled with care and not in organic compounds it is safe and we do pass through our bodies considerable quantities of mercury from various sources. Recently, accurate methods of determining the mercury content of animal tissue have been invented. Mercury compounds are used in fungicides in the Swedish timber industry and much of this enters rivers and lakes in which logs are floated, and is subsequently drained into the Baltic Sea. Excesses of mercury have been found in certain coastal fish, and they have been prohibited from sale for human consumption. It is not really known how poisonous they might be!

Mercury poisoning used to occur in babies who were treated with certain teething powders, and swallowed too much of it. Rather large amounts of it have been found in some of the larger fishes; this has been explained by the fact that they are at the end of the 'food chain' which means that successively larger animal species feed on smaller species and each concentrates more mercury in its body than those lower down. When it was discovered in a batch of tinned tuna caught off the US west coast, they were destroyed on the assumption that they were a danger to public health. Subsequently equal concentrations were found in a museum specimen half a century old, and it is probable that there is no real danger to human health. The study of mercury in animal tissue is a relatively new one and we need vigilance, but we need not be scared. Perhaps it is as well to take precautionary action at first, but the scare is a silly basis for permanent legislation.

The wide dispersion followed by subsequent concentration of a substance in the bodies of animals at the end of a food chain is one of considerable interest; it must be understood before rash action is taken. Mercury in large fish could be due to the food chain effect entirely. The amount put into the ocean by man is very small by global standards and is not altering the average concentration by measurable or significant amounts. On the other hand the mechanisms of dispersion in the ocean act quite slowly, and the concentrations in coastal waters or enclosed seas such as the Baltic may be rather high so that fish feeding there may be badly affected. How far Pacific Ocean fishes range is not known, and so there is much uncertainty.

The case of DDT is rather different because it has been found all over the world and yet it is not naturally produced. The fact that it has been found in Antarctic penguins is a tribute to the dispersive powers of the atmosphere as well as to the salesmanship of the manufacturers. We should be glad that it has gone so far because that means it has been well diluted. On the other hand, this rapid dispersion throughout the world does not mean that all the DDT ever produced has been equally widely dispersed. On the contrary, most of it must have come to the ground within a few miles of where it was sprayed, and so there is a vast amount still in the ground which will, in the course of the next decade or two, make its way into the sea.

Whether the penguins have eaten fish that have eaten smaller fish that . . . , which lived in the coastal waters of California, is not known: it is more likely that it has been dispersed by the wind to Antarctica and rained out there.

A complete ban on DDT has been demanded, but this is the reaction of a scared public. There is now a problem because a vast amount of DDT has been dispersed in the world. There has been some anxiety lest it should never be degraded and rendered ineffective in nature, but Soviet scientists claim recently to have discovered ways in which it is broken down. Likewise it has only recently been discovered that carbon monoxide is absorbed by some life forms: the fact that we did not know where it went to does not mean that it remains in the air, and we knew that CO must be absorbed or oxidised somewhere in nature because it was not accumulating. When we know more about the degradation of DDT we shall be able to make a good law about its control, and in the meantime its use should be restricted to cases known to be economical and beneficial. In Britain this is achieved by a 'voluntary ban' on its use.

The worst effects of DDT are not in the food chains at all, but in the insect world. Because of its persistence it kills a vast number of insects other than the intended victims. It, and other insecticides, have killed most of the bees in Britain with the result that honey can only be produced on a much reduced scale. Whether there is any effect on pollination remains to be seen.

The cause of the problem is the terrifying inefficiency of our methods of application. Only about one-hundred-millionth of what is used makes contact with the intended target in crop-spraying operations. The rest becomes undesirable environmental pollution. It is important to prohibit the dousing of crops from aircraft or tractor-drawn vehicles, but it is good to let the farmers of the Cheviot hills spray the backs of their sheep to kill the maggots. By that means a 20 per cent loss of sheep killed by maggots in hot summers has been completely prevented. With proper application no other insects are in danger, and the efficiency is raised from one in a hundred million to perhaps one in a thousand.

We may have to start thinking in terms of a labour intensive method, and improve that, rather than employ overkill methods. It is ludicrous to suggest to a cotton farmer in Sudan that he should have a team of men and women applying insecticide by hand to the part of the crop that is in danger, when all he now needs to treat several square miles of crop is one aircraft and a small land crew. But equally he may have to think of diversification of his crop simply because a complicated ecological system is much more stable and easy to manage than single-crop cultivation over a large area. When predators are no longer balanced with their food, insect problems can be created by single-crop cultivation.

All these scares are serious matters in the sense that we must learn quickly what effects we are producing and stop any bad effects quickly. But there is no doubt that they arise largely because humanity increases the scale of operations too quickly without really being in control of the consequences. The basic problem is the rapidly increased mechanisation of civilisation, and the increase in population which demands the development of single-crop cultivation on a grand scale. The probable solution is that agriculture will become much more variegated and labour intensive in the future.

An obstacle to good decision-making is disagreement among the 'experts' on technical matters. This does not arise because of genuine disagreement in most cases because, given the opportunity to discuss the question together, almost any group of scientists would come up with an agreed statement of the state of human knowledge on the subject and probably some recommendations based on their degree of certainty. The disagreements arise because the scientists are enrolled in opposing camps before they have had a chance to get together. Nor must we forget that it is a common failing to wish to be the author of a correct forecast, more especially when the forecast was of some sort of catastrophe; if the doom really was inevitable and one can be excused from the charge of not taking effective action to avoid it, there may be some satisfaction in having correctly given the first warning.

Tainted with these faults is the supersonic transport (SST) controversy. The argument is really about economics. Some say that if the SST were a profitable venture for humanity private capital would readily be subscribed for it and no government money would be necessary. That naïve argument, besides being rather obviously wrong in fact could be used against any major public expenditure such as roads and sewage treatment. But it is unsound in economics, too. If private capital is not subscribed for a venture it could be that the terms of profit offered are not as good as can be obtained for some other venture. We tend to think of money in terms of profits every year or half-year, and a promise of profit beginning in 15 or 20 years' time is not attractive nowadays. The assumption that the greatest public benefit comes from the investments which maximise the particular kinds of profit attractive to the particular class of person who has spare money to invest, or transfer wilfully from another venture, is a dogma which makes the stock exchange into a fatherly god.

Big money is involved in SSTs and so a grand issue exists long before the wretched scientist is invited by this side or that to give his evidence. The public mood is much affected by the widespread concern for the environment and so, to the opponents of the SST, a scientist who can predict a danger to the atmosphere or to Man through a modification of the atmosphere is worth far more than one who can find dangers for the crew, the passengers, or any other élitist minority directly concerned with the venture.

The issue has psychological overtones. Scientists are not used to taking sides dogmatically in other people's arguments, but they hold tenaciously to their assertions like a prisoner in the dock. It is therefore the first commitment to the cause that matters – a loud-mouthed scientist can make himself a committed, and perhaps eventually an embarrassing, ally. The controversy is bedevilled by journalists and TV companies who simplify the issue and mislead the public, partly because they do not understand, and partly because they are more interested in a circus performance than in the truth. The whole matter is treated like a legal battle between groups who have taken up committed positions, whereas it ought to be a colloquium of scientists all in search of the same truth.

Apart from the sonic boom, which is a quite separate issue from pollution, the SST is mechanically another aeroplane consuming more or less the same kind of fuel and producing the same kind of exhaust chemically, and using about the same amount per passenger mile (i.e. between one-fifth and one-tenth of the amount used per passenger mile by a large passenger ship, on an Atlantic crossing) as other jet aircraft now operating; and it presents much the same airport noise problem as a Boeing 707, or a VC 10. The only significant difference, apart from its cost and its speed, is that an SST flies in the stratosphere where none of the scavenging mechanisms which clean pollution from the lower part of the atmosphere operate. Consequently, it is necessary to investigate whether the accumulation of pollution over a long period might have undesirable consequences.

Before discussing the scientific issues it is relevant to mention a reserve argument which has been used by anti-SST lobbyists when it has seemed that the scientific argument about danger to the environment is going against them. Instead of reciting the case put forward by the scientists on their own side, and perhaps mentioning answers given to the case by the scientists opposing them, they withdraw from the scientific argument altogether and merely take a stand on the grounds that 'experts are in disagreement', and therefore it is necessary to pursue a cautious approach and act as if the worst forebodings were correct.

Several dangers have been considered. First was the possibility that the water vapour in SST exhaust might form contrails (exhaust condensation trails) such as we often see behind ordinary jet aircraft flying in the upper part of the troposphere. When SST operations became extensive, it was argued, these trails might form a layer of cloud which would cut off a large fraction of the sunshine reaching the ground and thereby alter the world climate.

It is customary among meteorologists to advise the air forces of the world that no visible trails are or can be formed in the stratosphere, and there is no reason to change this view. Although it is cold enough and although in some circumstances non-persistent trails are occasionally formed, there are no circumstances in which they could persist for more than a minute or two, and generally only for a few seconds. Such trails as do very rarely occur are only about half a mile or less in length and never become spread out, because the air is too dry.

Next it was argued that because of the low humidity of the stratosphere and the slowness of the mixing between it and the troposphere below, the water vapour in the exhaust would gradually raise the humidity to a new equilibrium level at which the supply from the aircraft equalled the rate at which the water vapour they introduced was carried down into the troposphere by the natural movements of the air. The consequence of this would be that the exchanges of heat between the stratosphere and the air below and outer space above would be upset. The effect would be to reduce the escape of radiation from the lower part of the atmosphere and thereby make it warmer. This is the well-known 'greenhouse effect', and climatologists have long argued about how changes in the composition of the atmosphere might change the temperature and perhaps be responsible for the ice ages of the past. It is generally agreed that dust put into the atmosphere by volcanoes reduced the input of sunshine and was a contributory cause of ice ages. Out of this arose the idea that an increase in carbon dioxide would cause a warming by increasing the greenhouse effect: undoubtedly if the CO_2 in the atmosphere were substantially increased a warming would occur and the only question is how much. The original scare was that the CO_2 put in by burning fossil fuels (coal and oil) would warm the air so much that the ice of Greenland and the Antarctic would be melted and the sea level raised so much that most of the world's largest cities would be flooded by the sea. It is now thought that the uptake of CO_2 by vegetation and by the sea would prevent the CO_2 in the atmosphere from ever becoming enough for this to happen. Likewise calculations show that the effect of increasing the water vapour in the stratosphere would be negligible.

But water vapour (H_2O) can have other effects. In sunshine many chemical reactions take place between the oxygen, nitrogen, and water vapour in the air, and consequently a multitude of their compounds is present. The sunshine falling on the upper layers of the stratosphere, and in particular on the top two-hundredths of the atmosphere, dissociates the oxygen molecules at the very low pressure which obtains there into atomic oxygen atoms (O). These tend to form ozone (O_3) molecules by reaction with molecular oxygen (O_2). The ultra-violet (UV) component of sunshine now falls on the O_3 which is dissociated back into O_2 and O, and at the same time absorbs the UV radiation. From the point of view of life on earth this is very important because if the UV radiation reached us it would cause a blistering of the skin, possibly blindness, and other debilities. Life has evolved protected from UV radiation by the O_3 and it is important that it should remain there.

The picture of the chemical reactions just given is very much simplified. In fact there are very many reactions which occur in sunlight which involve H_2O, nitrogen (N_2), oxygen (O_2), and various oxides of nitrogen collectively represented by the formula NO_x. The amounts of NO_x and H_2O present partly determines the amount of O_3 present because they react with it, and there is a series of reactions in each case which together reduce the amount of O_3 without altering the amount of NO_x and H_2O. The O_3 is ultimately turned into O_2 which does not matter because there is so much there already, but a reduction in the amount of O_3 alters the amount of UV radiation absorbed and so, it was argued, there is a danger that the amount of dangerous radiation reaching ground level would be increased.

Now it so happens that in absorbing the UV radiation the O_3 is warmed, and so the absorption is very important in determining the temperature of the higher parts of the stratosphere. Indeed, from measurements of the temperature the amount of O_3 can be deduced, and from that the amount of NO_x can also be deduced because more or less of it would alter the amount of O_3. The calculations show that there is already so much NO_x in the stratosphere that any put there by SSTs would have a very small effect: the more there is there already, the less effect a little more would have.

In conclusion, therefore, it can be said with considerable confidence, that the effect of having a fleet of a few hundred SSTs flying in the stratosphere would, at worst, cause a warming of the atmosphere of the order of $\frac{1}{2}$°C. This, of course, is a big effect by human standards, but it is small compared with the effect of changes in the total CO_2 content of the atmosphere since the Industrial Revolution began; less than the effect of increased dust and smoke in the atmosphere due to tropical agriculture (which makes the atmosphere colder); less than the effects of man's activities of the more distant past such as extending the deserts of Africa, the Middle East and Asia by overgrazing, and thereby causing more dust to be blown up into the atmosphere; and very much less than changes in air temperature due to natural causes such as volcanic activity and others, as yet only hypothesised, such as the opening of the Bering Strait and a consequent rising in the temperature of the Arctic Ocean.

All these arguments arose from attempts to influence the decision of the US Government about whether to subsidise an SST or not, by making people scared of possible ecological consequences. It is worth considering whether the judgments just indicated depend critically on very complicated calculations of the amounts of various chemical compounds of the common elements that are present in the stratosphere. Can the ordinary man have any means to judge for himself in these matters? It is not enough to examine possible ulterior motivations for the statements of scientists. To what extent can it be seen that many of the scares really belong to the realm of science fiction?

Perhaps the approach is best indicated by saying that almost everything that could happen to the common elements in the atmosphere actually does happen, and quite extensively, in nature. The original scare about the O_3 being depleted by putting NO_2 into the stratosphere was founded on the assumption that there was almost none there already. But oxides of nitrogen are formed everywhere the temperature or pressure of the air is sufficiently increased. It happens in internal combustion engines. It happens in volcanic eruptions, in lightning, and in meteor trails (shooting stars). Volcanic eruptions often send great masses of hot air from the troposphere into the stratosphere. A moderate eruption would be equivalent to perhaps 100 SSTs flying for up to a year in this respect. Lightning, which supplies most of the NO_x in the troposphere (and most of what would be carried up in the plume from a volcano), puts in about as much as 10,000 aircraft flying all the time, and meteors supply it to the very high parts of the atmosphere above most of the ozone. Probably the NO_x is converted into nitric acid photochemically in the presence of water vapour.

The atmosphere is in a continual state of turmoil, so that it also contains other compounds such as SO_2, SO_3, H_2S, NH_3, and CH_4 which are emitted at the ground and which are all chemically very active, and all of which (particularly SO_3 and CH_4) are found in the stratosphere. The atmosphere is a very inefficient engine, which cannot easily be disturbed from its course. Only in regions which are stagnant, such as the air over Los Angeles, does the accumulation of pollution produce significant effects and then only locally. The whole atmosphere is pollution: all its components are continually being renewed and removed, and intruding substances are quickly dealt with in the troposphere, and slowly in the stratosphere.

In addition to the reasons already given – which amount to saying that aircraft exhaust consists of materials that are plentiful in nature, and that a bit more will not make any difference – there is a fundamental fallacy in the supposition that even if it did have some magic effect on the ozone in the stratosphere, aircraft exhaust could have a significant effect. The only reason why there could be any danger is that the mixing mechanisms are very weak in the stratosphere. Otherwise it would have exactly the same composition as the troposphere. In order to have any effect on the chemistry of the stratosphere it would be necessary for the exhaust to be mixed into the whole body of the air. In fact, as can be seen from visible exhaust trails, only a small tube of air is affected (when aircraft trails seem to cover the whole sky it is because this tube has been drawn out into a very thin sheet of the same volume as the tube), and at the worst only thin laminas of air could be affected, with most of the air in between behaving as it always does. You cannot have it both ways! Either there is very little mixing in which case most of the air gets no exhaust at all, or there is a lot and the air is like troposphere air and there will be no effect. The theory that the ozone would be significantly depleted as a result of exhaust being mixed into it was put forward by a Californian chemist who was used to conditions of the laboratory and the stagnant air of the Los Angeles smog. His conception of the motions in the stratosphere was utterly naïve.

The only possible lasting effect Man could have would be by putting enormous amounts of dust into the stratosphere: this might perhaps produce an ice age, and once the dust was there it would take a long time for it to be removed naturally and there would be no hope of doing so artificially. Such an effect could only be produced by indiscriminate nuclear war.

The atmosphere is not susceptible to catalystic effects because nature produces so many substances that there is no plentiful one which it could make much difference to add; they are all already there. Even the hopes of producing rain artificially have been utterly disappointing for this reason: clouds do react a little to an artificial stimulus, but the stimulant gets washed out in an hour or two so that only by a continuous supply far beyond our means can continual effects be achieved.

We have good knowledge of the dire consequences of continuing to do many things we are already doing and it is much more important to find ways of stopping them than to avoid starting certain enterprises on a comparatively small scale. We ought to keep our attention on smoking, and on nuclear weapons which cannot possibly be used for a purpose which would subsequently be approved of. Their continued production is contrary to all the interests of humanity because they could not all be used without terrible damage to civilisation far beyond any possible benefit. To these two obvious evils must be added the pollution of the sea and inland waters, and excessive population growth. These latter two problems really are scaring to think about seriously.

One of the dangers of scaring people is that they become more inclined to think in terms of prohibitive legislation enforced by penalties, and that way lies social disaster if we do not reduce the economic incentive to making pollution. If the penalties, and the possible gains, are raised high enough we are asking for criminal organisations to take over the economy. Scaring the people produces irrational, emotional reactions which are inefficient and play into the hands of the demagogue.

The atmosphere, and indeed the whole world, is very complex. The scaring themes are usually oversimplifications. Public clamour is needed in order to get action, and our dilemma is not to use scaring themes to raise that clamour. If we cry 'Wolf!' or 'Fire!' and there really was none the public will only believe the evidence before their eyes. They will ignore the pollution, distress, and exhaustion of resources which our way of life causes in other parts of the world. Before long our sources of raw materials will either become too expensive, or they will become exhausted, or the local population will rise in rage at our indifference to their fate and the resources will be cut off. Then it will be our turn to cry 'Help!'.

Is international agreement necessary?

The advance of 'Industrial Revolution Man' has been achieved through international trade, and any attempt to change his way of life will be eased if its programme is agreed internationally. Such agreement is in the interests particularly of the rich countries because they have most at stake, but it means agreement not only among the rich.

Agreement is not easily come by, such is the tradition of self-interest among all governments; and so it must be worked for hard and long. The most serious danger was hinted at in the final paragraph of the last section: the people of the poor countries will one day rise in rage if nothing is done to fulfil their aspirations. A microcosm of the economic apartheid which will face the world if the rich continue their path of nationalistic economic growth is already in existence in South Africa. There, there is no question of sharing: the haves represent their position as the achievement of their own civilisation, the work of their own hands, their culture which it is their duty to preserve. Apartheid is maintained by force of arms which supports the economic and legalistic system which is the way of life.

Already there are more starving people in the world than

there were people in the whole world at the time Malthus predicted that population would run ahead of food resources. Because the whole world is not starving uniformly it has often been argued that his prediction was wrong. The difference in the world today from the world of the time of Malthus is that all people are conscious of their rights, and are claiming them. One symptom of the age is the attempt to dispossess the Australian Aborigines of land assigned to them in perpetuity in the Northern Territory. In this land mineral resources have been discovered which it would be very convenient for Japan to exploit since its economy has outstripped the resources of Japan itself. The argument is that the Aborigines would make no proper use of the land, and that it would be better for the prosperity of the world if it were 'developed' by 'Industrial Revolution Man'.

Only international pressure to force higher moral standards on the nations of the world can avoid conflict. The technocrats of the rich countries can see in their minds a way of solving our pollution and population problems and of providing enough food for us all; but their vision has no place for the arbitrary behaviour which we call freedom, for individual pride and dignity. As bosses they cannot appreciate that their vision is a slave world, with apartheid on geographical (and, in practice, probably racist) lines. The have-nots of the world want dignity on their own terms, not prosperity on terms dictated by the haves. A mere forty years ago the haves and the have-nots were defined by Europeans according to colonialist and imperialist concepts; the have-nots were the European countries which had no colonies. Now imperialism is dead and has been replaced by racist concepts in the minds of those who do not see that in a full world no one can be kept ignorant for long, nor kept in slavery.

The Americans, Europeans, Japanese, and their ilk in other parts of the world, consume and create pollution per person at twenty to fifty times the rate of the less industrially developed countries – that is those in which the horsepower generated per person is much less. If all the world's population lived like the rich there would be imminent ecological disaster. It will not happen like that but, if there is to be justice not only as between the members of one nation but also as between the peoples of the world, the rich must soon put their economies into reverse in the sense that they must consume less and less tonnage of materials.

It is urgent, therefore, that the international trade of 'Industrial Revolution Man' be quickly reorientated towards conservation of resources and reduction of pollution and towards an equalisation of wealth in the world. There is no justification for further advances in personal acquisitiveness. The race to achieve the ultimate in consumption has been won, and it is foolish to continue to race beyond the winning post.

Industrialists of today are under great pressures to behave in an anti-humane manner. Take-overs and rationalisation of businesses require that factories be closed and work be automated. With our excess of cheap fuel our system cannot be bothered with human labour; the returns are much greater from employing machines. Like army commanders attempting vainly to minimise casualties the captains of industry are unable always to abandon their objectives on the grounds that the human cost is too great. The unemployed are not their personal concern: however much they are the concern of society as a whole, including the captains, they are not the concern of the captains in their jobs. They have to face what they call 'the facts of life' by which they mean that they would go out of business and lose all influence in society if they allowed their compassion to dominate their business acumen.

They reckon, also, that their business activities serve their countries well, and indeed by some criteria this is beyond question. But what a world of contrasts and calamity they compel the beneficiaries of their exercises to live in! They are under no compulsion to seek the welfare of the have-nots.

All professional groups in commercially advanced countries are unable to give up any advantage for their members which they feel able to grasp. However much a doctor or engineer or professor feels that he and his fellows have as much affluence as they need, as they want, and as they ought to have, they cannot get their profession as a whole to refuse any further increase of salaries. It is as difficult for a rich man to enter the kingdom of heaven as for a camel to go through the eye of a needle!

This difficulty facing intelligent and perceptive people

who happen to belong to a privileged group is almost universal. They are unable to follow Gandhi and wear poor men's rags. It is difficult for a white man in a country in which other races are discriminated against to make a black man feel him as a real friend of the heart. An intellectual friend perhaps, but he goes home to his privileged household every night. Yet the alternative to recognising that they have already arrived at the best affluence the world can give, and that all their problems of living lie outside the material sphere, is to seek more affluence in the silly hope it will solve those problems, and keep affluence from others by force.

Governments see themselves driven by the conventional measures of success and keep on racing well past the chequered flag because they know no other way. Just as the manager needs government to give him the power and incentive to serve his fellows well, so governments need international agreements to change the man-made criteria which they see as 'facts of life'.

The revolutionary misses his chance if he vilifies the manager for indifference to the welfare of his fellows (labour force): he should be seeking to change the measure which forces the manager's hand.

If consciousness means anything it is the power to take hold of our destiny. Our industry progresses with no guiding hand laying the rails on which it must travel without thought of the destination. If a god once ruled, it is now clear that he abdicated long ago. There is no natural law for 'Industrial Revolution Man' in a full world: it is our responsibility consciously to create it.

6 The Cultivation of the World

Was Malthus right after all?

The famous Dr Malthus thought that population would multiply faster than food production could grow. He was not primarily concerned with the finite size of the world but that, by increasing more rapidly than food supplies, the population would condemn itself to starvation. It is often said that his gloomy prediction was wrong; but was it? Today there are more undernourished people in the world than there were people in the world altogether in Malthus's day, and unless we are ready to magnify the misery and create perpetual injustices, the growth of population must be urgently restrained. This is so close to the heart of our problem that in discussing pollution, and air pollution in particular, we must accept it as a starting point. No arguments about it will be entered into here except to say that it is no solution to show that a certain very large population, of say, ten thousand million, could be maintained on the earth. People will not tolerate indefinitely a kind of battery-farm existence, with all chemical necessities provided by benevolent technocrats. That is no one's ideal.

It really makes no difference to us whether Malthus was right or not. There is a danger, however, that in trying to discredit some other prediction of doom it will be said 'these predictions of doom are always wrong – Malthus for instance'. An instance does not prove a generality, but in this case the instance is very questionable anyway. There is yet time for Malthus to be right, even if we do not accept the argument just given that in a sense he is right already.

Some gloomy predictions are obviously wrong. If the figures of the amount of horse manure taken from the streets of London in 1800 and in 1900 (roughly 300 tons a day) had been extrapolated exponentially it might well have been shown that the whole of London would be 1,000 feet deep in horse dung by the year 2000 unless the railways did nothing but carry horse dung to the country all day and all night.

But some predictions come too late! The buffaloes of North America numbered perhaps hundreds of millions; nobody will ever know how many there were 200 years ago. Then with horses and guns they were rounded up and slaughtered as if there was no end to them; and suddenly, there were none left. Likewise the carrier pigeon of North America used to be everywhere; a pest. But ruthless shooting almost suddenly caused it to disappear. Now it is extinct. The white rhinoceros numbers probably less than 100, certainly less than 1,000 in the whole world, and it was only through the hard work of people connected with the zoos of the world that it still lives, however precariously.

The world is full of Man. Evolution is now proceeding along a new path; unless great care is exercised the number of species, instead of continually increasing, will catastrophically decrease and a biological heritage created over aeons will be thrown away for ever. This is not true simply of animals; vegetation is in a similar danger through mindless one-crop cultivation in which all other (interfering) species are destroyed as weeds.

There is no safety in large numbers: the multi-millions of 'locust birds', the red-billed quelea, may one day suffer extinction at the hand of Man, like the carrier pigeon, because of the damage to crops in West Africa: its ravages can cause widespread human starvation. There is no safety in small numbers: the Bengal tiger competes with Man ever more ruthlessly for the unattended countryside. There is no safety in remoteness: the reindeer of Lapland will suffer the fate of the North American buffalo unless Man deliberately holds back and does not yield to the pressures and incentives that have driven him on and on till now.

Who is this Man?

We speak of Man as if he were a homogeneous species, possessed of a single character the world over. The discovery of fuel and how to use it in the temperate climates of Europe allowed

9

one branch of Cromagnon man, thousands of years after he had slain the last more primitive, Neanderthal man, to satisfy his zeal for conquering, for storing, for planning, for managing his survival through the cruel winter by creating the material comforts of civilisation. He cultivated his continent, became imperialist, and because of his superior military power and organisation, never questioned the assumption that all men throughout the world would accept the European criteria, and willingly join in the grand advance of progress.

What has this purposeful creative Man become? He carries his Christianity and his technology across the world expecting to be thanked for both, only to find that he has stimulated competition and an exacerbation of all those traits of intolerance and arrogance which he is least proud of but best exemplifies. Perhaps the most ludicrous of all incidents in the evolution of ideas about the enslaved and exploited peoples of the world was the cynical admiration which the white man offered to Martin Luther King after his assassination. They saw in him a man who had learned his Christianity well: 'If only the others would do likewise,' said the white man's leaders, 'we could lead you all into a new era of peace, co-operation and prosperity.'

If the non-Europeans of the world would be content with a view dominated by the backs of the Europeans they are following, all would be well for the technocrats. It so happens that not even Europeans are content with their appointed role of managers of the factories, factory farms, factory fisheries, and factory teaching institutions of the world: from their own university production lines come some of the most vociferous heretics, who receive the appellation 'anarchist' because they wish to upset the orderliness of capitalism. They cannot any longer be bribed to run the system: it is managed more each day through fear, through desperation; through inability to see any alternative. It is the managers who fear and despair.

In searching for a way of life in which Man's influence will pervade every corner of the world we must not think in terms of the technical aspects of cultivating the whole world – either as a factory farm or as a garden of Eden. The spirit of Man is not simply the spirit of 'Industrial Revolution Man'. It is the aspiration of us all.

Above everything else we must plan to have room to

manœuvre. This means space for experiment, time to think, and opportunity for evolution to operate. Somewhere for the weeds to grow, a place not dominated by a single-minded purpose. Our most difficult task, after all the generations of greed and competition for power, is to plan to let part of the world be unplanned, where Man can look for inspiration in the way his dreams have been influenced by thoughts of the inaccessible cosmos.

The atmosphere above all

The atmosphere alone destroys its blemishes. Because of its power to cleanse itself it seems likely to remain the last earthly stronghold of nature uncontrolled by Man. In this there is a lesson for Man, the gardener of the world. We need to leave a vast expanse of ocean, or fields untilled for a season, where the biosphere can heal the damage due to industrial activity. In a world full of Man, by-products put out by Man will be distributed more widely by the atmosphere than in any other way, and it will be necessary to use the atmosphere for this purpose to an ever-greater extent. For example, incineration of dangerous material at sea will enable it to be absorbed harmlessly into the biosphere more readily than dumping at sea where it may 'commit a local murder' of which the corpse would remain a record of Man's carelessness for a long time.

But if we fail, if Man drifts on, forced at last to mend his ways by a succession of environmental disasters on land and in the water, still the weather will remain for the most part quite unaltered, a continuing source of delight and variety to elaborate and welcome the human experience.

What men believe about themselves

There is no doubt that in a single community there exist people whose concepts of the world and its possibilities are totally opposed to each other. On the one hand the engineer-business man: he has made a great contribution to human welfare and prosperity. He is very proud of his achievements, and he believes the limit of the kind of material progress he has pursued to be still very far in the future. The markets can expand until all humanity can enjoy the benefits of modern technology.

Nevertheless, he is very conscious of the existence of other men in his own world. He has continually to deal with trade union representatives whose view of life is that of the average employee. The ordinary member of the labour force is a realist: he knows that he and his mates can only achieve a level of sumptuousness that can already be envisaged; he limits his ambitions by what he can see to be practicable for the ordinary working man, and mostly he and his mates expect to remain ordinary working men. Their ambition is a level of quietude that can only be achieved by some sort of protection against the whims of his employer and the vicissitudes of the market. If a man chooses to set himself apart from the masses and become a boss he cannot expect to share other men's objectives; his values are not appropriate except for a minority and for a society which tolerates such a minority. He cannot complain, indeed he should be thankful, that all men do not share his ambitions, and in so far as he sets himself up as his own ideal he must recognise that most men will have other ideals. For most men there is something which we call enough, and they expect and hope for no more. The few aggressive Christian technocrats have invented an ideal man – hardworking, efficient, self-reliant. But he has none of the peasant virtues: he only savours life deliberately and at the appropriate time. His hey-day is passing as the Industrial Revolution mounts to its crescendo, and already his children cannot communicate with him as he did with his father.

Growing up is a new generation with an attitude to affluence, aggressiveness, and acquisitiveness nurtured in different surroundings, aching with needs unknown to most of its parent generation. The greatest mistake we could make would be to plan the cultivation of the world on the assumption that we, now, fostered in the middle of the twentieth century, could envisage an enduring ideal for humanity. For this reason alone we must retain room for manœuvre while we can get it.

Blueprints for posterity will be argued about at great length but stimulating as they may be for our imagination, as mirrors in which to assess anew our own civilisation, they can never suffice for actual posterity.

What is this room for manœuvre? It is the opportunity to choose, to experiment with the organisation of society, to remain uncommitted to much of our own future. It means not only having some wealth in hand, some land to spare but also a versatile political structure which allows for experimentation rather than the single-minded pursuit of ideals and the ruthless application of doctrine. It is more than providing for pragmatism; it requires a certain deliberate inefficiency, a many-mindedness.

The successful businessman complains of the disappearance of the craftsman, yet it is his own single-mindedness that has chiefly contributed to this regrettable situation. His measures of success and efficiency are out of date for the next stage of human evolution – the completion of the occupation of the world by 'Industrial Revolution Man'.

Until now he has been concerned mainly with production. From now on let his motto be *Nature makes no distinction between products and by-products*.

7 Meanwhile, Back at the Ranch . . .

To be or not to be

We are the products of our history. We have reasons to hate capitalism but we love our freedom. We enjoy immensely the opportunity that we have as individuals to pursue our own objectives. We have so much wealth that we do not need the terrible repressive laws against theft that were normal only a century or two ago. Either we have a religion which allows us direct access to God so that we can say 'boo' to the priest, or we have no religion and recognise the authority of no man as absolute.

This very freedom makes us feel the responsibility individually for doing something effective. Much as some people may regret the anarchic state of our politics, it is the realisation of the ideal of freedom for each of us that we are witnessing – only in the wealthy societies, however. Are we to let domestic extravagance rip or is there hope through individual action to restrain it?

Progress solves few problems; it simply makes old ones irrelevant by changing the scene and this, of course, brings new ones. In particular, it shortens our vision of the future and by the complexity of industrial civilisation makes it difficult to see what an individual can do by his own choice that is effective. The rush to more rapid exploitation of natural resources and the greater dividends that brings makes us feel caught, as if the rest of society pursuing its acquisitive objectives will pass us by if we stop to question whether we are travelling in the most desirable direction.

The assumptions of our growth-orientated society are so well educated into us that it keeps being said that we cannot understand destructive tendencies in modern youth. They ought to be quite obvious, because there is nothing natural in a well-ordered society in which we all play our role of good customers and hard workers at the behest of the technocrats. To what end do we operate our machines if we are told how to behave in our freedom?

There is a great temptation to abdicate from individual responsibility and leave it to the government or to God. If we succumb, we choose 'not to be'. We are a mixture of peasant – who wants to savour the mysteries of life, the sky, the taste of fresh food, the noises of nature, the emotions of human relationships and song and contest – and of explorer and creator of new things and new experiences. In particular, we want to be a major influence in the creation of our own individual environment: to mark out our territory, furnish our own homes and keep the door locked when we choose. The revulsion of modern youth is very rightly against being told precisely how to do this. Do not the technocrats understand that doing a thing is more important than ensuring that it is done the best way? Especially when the criteria of what is best always turns out to be arbitrary.

As the world becomes full we see the environment being enclosed all around us: the efficient use of land, minerals, anything, is deemed to justify its destruction as a realm for dreaming, for living, for wonder and for personal renewal. The whole economy is measured by scales which ignore these things and when this is pointed out the economists reply by trying to measure them. But we must not accept the contest under such false rules for that would be equivalent to begging the whole question.

We want to be. We can see the trends going in a direction which will destroy for posterity the personal advantages which the Industrial Revolution has won for so many of us. We want our children to be. We are faced with the problem of changing the deliberate purposes of much of our system of accounting and of incentives, and this is a greater task than the one faced by Marx. We must decide whether we shall attempt a gradual change, or shall agree with revolutionary youth that there is no hope in gradualism and the system must be destroyed before we can build up a new one.

The cul-de-sac and the trailer

The world is finite. We are travelling down a cul-de-sac,

enjoying the travelling but scarcely aware that our source of pleasure is soon to end and that we have not in the least thought out what we are going to do instead of indefinitely increasing our population and our ravages of nature.

We are on a vehicle which draws behind it a trailer called unemployment. What is worse, the trailer is attached by means of an elastic rope called inflation. In our enjoyment of the ride we have stretched the rope, and the trailer therefore threatens to catch us up and crash into us from behind. Our conventional reaction is to accelerate still more and pursue ever greater economic growth. Obviously this merely stretches the rope still tighter, storing up more energy for the crash when the world can accommodate no more expansion of humanity.

The ride is fun, and our drivers keep telling us that we have always kept the trailer at a safe distance in the past and by the same methods can obviously continue to do so in the future.

If we cut the rope and pursue a kind of industrial apartheid enjoying the ride without the trailer, ignoring the Third World, entering into a self-sufficient Europe, this will not lengthen the street, nor will it prevent the eventual crash of the free-wheeling trailer into our behind. Furthermore, some of our fuel supplies are on the trailer.

If we postpone decision not only will the crash into the wall at the end of the street be more disastrous, but any further momentum we now give to the trailer will worsen the impact upon us, and that impact will take place at the moment when we are already stunned by the crash. There is a terrible moral problem in feeding the hungry, poverty-stricken world in circumstances which only encourage population growth and make the prospect for immediate future generations yet more grim.

A calamity of major proportions is inevitable as a result of the population pressures and the unemployment which we are creating in industrialised societies. Unemployment becomes far worse in the unindustrialised world as we try to save them by industrialisation. We act as if they will be able to sidestep all the horrors of industrialisation which our forebears went through; actually their predicament is more serious because the population growth has preceded, not accompanied, the industrialisation.

Clearly our task is to start the deceleration now to avoid completely the crash into the end of the street or, to use an earlier terminology, to give us room for manœuvre, to retain for our children some possibility of choice about what kind of living they want. At the same time we must plan to absorb the impact of the trailer.

There is no possible benefit, except in a gluttonous sort of way for the present-day rich, in establishing an economic apartheid. The world must develop as one, now. But only if we plan ever more effectively to stop population increasing to the limit of resources can we hope for the continued benefit of cultural variety. We can hope only for two or three generations of terrible conflict if we let population extend itself beyond the world's limits at present living standards.

So ingrained are our attitudes, and so well do the pressures maintain them, that the best real hope lies in the young. Fortunately there is a very rebellious attitude in educational institutions and it is of great importance that we should cease for ever to regard our schools as places where we inculcate attitudes of compliance. New objectives must enter the hearts of the peoples of the rich countries.

New criteria

In a global scene of such dramatic magnitude the personal effort to save fuel, salvage rubbish, shun activities which make pollution, and plan our families, may seem of little consequence. By becoming aware of our individual powerlessness we see that action must be taken at governmental level.

When a dairy declares that 'for reasons of economy' it will sell its milk at a higher price in non-returnable bottles we know that the rules of accounting under which the dairy operates have become foolish. It no longer measures public benefit, and it may even be true that if everyone appears to benefit financially the whole of society may nevertheless be worse off.

If we were able to improve the efficiency of our use of fuel so that we used only half the present amount but continued to do all that we do at present except that everyone in the fuel business did half as much work, it would be measured as a bad thing unless we all paid twice the price for the fuel so that no one would have a smaller income. Whatever else our present criteria might make of such a situation, conservation of world resources for future generations would not appear in the accounts. Nor would the accounts contain reference to the reduction in noise and pollution which would result from the greater efficiency. In fact, as soon as the use of fuel became more efficient the present system would deliberately seek ways in which to continue to use as much as before. On its present criteria it would be unable to adjust to the idea of reducing fuel usage.

It is no use expecting accountants to apply idealistic judgments. It is not the accountants but the owners of wealth who become philanthropic when they have a great surplus to dispose of. It appears that when there is a surplus governments think only of handing it out so that people shall have more than enough: it is never allocated deliberately for purposes which seem remote – across the sea or in the future.

For these reasons a public opinion must be generated which will make governments act in the interests of the world community and posterity. It is unlikely that we can find a good substitute for money and the efficient way in which its distribution produces effects, good and bad, and so in the end we shall have to change the tax burden so that it compels the desired results. Taxes are arbitrary transfers of money designed to make the accounts serve a desired end.

The magnitude of the changes needed is not well appreciated at present, not even among people who ought to be well informed. Industrialists, economists, politicians, and housewives have lived at a level of material security that has only been temporarily threatened by war, and it is difficult for them to believe that the physical limitation of the world will soon be an absolute obstacle to further progress. Unfortunately discussions tend to be concerned with solutions of the problem. When we say that the *tonnage* of material consumption must be reduced in the rich countries and then find that the means needed to achieve that reduction threaten established practices which seem to be integral parts of our security and would seem to create worse problems of unemployment, we find ourselves caught up in arguments about these important details. Perhaps we do not yet have a solution to this problem, and when a proposed solution is found unacceptable it seems that the discussion points to going on as we are.

This is equivalent to saying that because we have not proposed an acceptable solution the problem does not exist. We should be saying that the solution must be imposed, or made acceptable, or a different one found.

To make new criteria acceptable we have to build up a consensus about what problems are urgent. Every political party will then be forced to have its version of a solution. The solutions must be primarily social and political. Up to now solutions have been thought of as primarily technological.

Thus, in the middle of the nineteenth century the River Thames in London was so foul because all the urban sewage was put into it, that muslin soaked in disinfectant had to be draped over the windows of the Houses of Parliament to make life bearable inside in summer. The construction of a proper system of sewers and sewage-treatment plant was undertaken at great public expense, but no one thought that we needed a change in the political system or a major change in the incentives offered by our culture. The problem was big but a basically simple engineering solution was sought and the cost found. London is now several times larger but there is no unpleasant atmosphere or health hazard from sewage.

We derive a feeling of security from such technological achievements, but we are now beginning to reach absolute limits in the use of water. There will have to be big changes from the present, essentially Victorian, sewage treatment and disposal methods if the population and its industry grow further in England. In many parts of the US the ground is becoming saturated by the infiltration of domestic sewage; and cultivation, health, and water supplies are threatened in many suburban areas. These areas are spread out because they have grown in the age of the car and any public sewage system they try to create will be very much more expensive and fuel dependent than in the more compact cities of Europe.

In this kind of light the following suggestions must be judged. There ought to be no conflict between those who try to dream up a new social order to meet the new needs and those who try to change for the better, by gradualism, the habits which seem to be getting us into difficulties. Above all, it is no solution to suppose that the rest of the world will automatically accept a solution that appeals to us. Indeed that should not even be one of our objectives. The loveliness of the world comes from its variegated cultures as much as from the variegated environment which produced them, and this is a most important reason why we should try to ensure that we have room for manœuvre. Otherwise the inexorable pressures will squeeze out variety of humanity just as it is destroying many beautiful animal species through the utter indifference of the competitive users of fuel power.

It serves no purpose to argue that the ruthless operators of our system must be stopped: the system makes them ruthless, it is one of the corrupting aspects of power. The system promotes those who are successfully ruthless, it seeks out and rewards the ruthless. It is no use getting angry with people who manage a system which promoted them; we need rather to see what it has done to them.

But generalities about the system are not likely to convince enough people to force a major political change because the system, with all its faults, has made people healthier and freer – and quite obviously so. In the world today it is certainly alleged and widely believed that everyone wants to have more and consume more. In our wiser moments we perceive that all is not well and that very soon all the evils of overcrowding and pollution will repress us. We know this not because of a general deterioration of our environment but because of the increasing frequency of incidents.

The incidents are difficult for a non-scientist to understand and there is a danger that excuses will be hidden away in the discussion of details. If we try to counter the argument by ourselves mastering the science, we can quite easily become bogged down in comprehension of the technology and in wondering whether or not technologists ought to give way to economic pressures in individual cases. We then lose sight of our objective of changing the trends of society.

If we decide on the other hand that it is now our turn to be ruthless, we are in danger of making decisions which are technologically unwise, and which leave many tendencies unchecked. Somehow we need to establish a more honest dialogue between those who explain and those who claim the right to decide democratically. The escape from commercial commitments which cause continuing pollution will not be painless: the technological problems are more difficult to understand but much easier to overcome.

A summary of objectives

1. In our personal lives we need more correct information about which of our activities is causing environmental damage. We must distinguish between different kinds of undesirable activities. Bonfires, crude advertisements, and aircraft noise are temporarily harmful to the quality of life. Building programmes which may dominate our cities for centuries, industrial investments in cars and motorways, thermal power stations, food manufacture, supermarket shopping, and power-driven agriculture could commit us to a way of life which we would find very hard to give up in the interest of human justice, while to continue in it would mean apartheid enforced by the rich.

2. Consumer organisations whose main objective has been to ensure that the buyer gets a good buy must now turn their attention to ensuring that society gets a good bargain in the long run. Pressure groups have always stimulated political and governmental activity when they have had a good case. But vigilance must be eternal: in penal reform and housing management we have made steps forward followed by retrogression due to public complacency. We cannot afford environmental complacency ever again.

3. Alongside pressure groups for the protection of our land we need industrial pressure groups. Duty to shareholders has often been regarded as paramount in the management of companies, and duty to the ratepayers in the conduct of local government. It has been accepted that ownership or payment confers a right measured by the cash involved: posterity owns no shares now, and many people have to pay for decisions with their livelihoods or forfeiture of amenities. Decisions of public companies must be watched and judged by wider criteria in the way the running down of the coal industry in Britain was designed to minimise social ill consequences.

4. Criminal activities, such as the illegal and dangerous

dumping of waste chemicals, must be treated as a problem of our society rather than as a crime. We must have a public disposal service or else nationalise the industry producing toxic waste, so that private profit is not a motivation for surreptitious dumping. In a free society we must seek this kind of solution rather than a penal one.

The structure of society should not be designed to lead us into temptation. If illegal dumping of waste chemicals remains profitable, merely to raise the stakes is dangerous; for sufficient profit a man scared to dump his own chemicals will call in the Mafia to do it for him.

An activity is only stopped by threat of punishment if it is both easy and no less profitable to live without doing it.

5. Certain values and practices are in each age regarded as obscene and pernicious. They are assessed according to how much they threaten the structure which society thinks is essential to its welfare. We have lately become much more permissive, perhaps because we are becoming more wealthy; and we shall find that personal greed and successful acquisitiveness beyond certain bounds will gradually be seen more as an anti-social quality as well as in the traditional moral light as selfish.

Equally, the presence of poverty in our midst will at last be regarded as a crime of neglect, not as a punishment for sloth.

A rich society must treat all its citizens with as much generosity as a rich man treats his own children, however unworthy. To change the mood of society so that a much smaller range of incomes is acceptable can only come about through continual activity at the level of small groups of people talking about their own and other people's status, developing new sympathies and antipathies which alter our basic motivations and status scales.

We need to reduce the range of incomes so that the richest has, say, only five times the poorest. If we were to plan a utopia we would surely have a range of perhaps only two or three, and so five is a modest objective. The vast majority of society already lives within that range and it would not require any anti-democratic development for them to impose it on the outsiders.

When that is achieved we can think of reducing the range further; then we shall be better equipped to help reduce the even greater inequalities between nations. At present we seem to be quite unable to make any contribution in that direction.

(The US aid given to India in twenty years is equal to the cost of one month of the Vietnam war.)

6. This last proposal would be facilitated by an alteration of income tax: it could be completely abolished within the agreed range of incomes and raised in a few years to 100 per cent for incomes above the range.

Revenue should then be levied on the real source of our present wealth – the consumed mineral resources, fuel in particular. This would make us give much more attention to avoiding waste of them, instead of wasting people, as we do now through massive unemployment, simply because fuel is ridiculously cheap. Ridiculously, because it is expended capital; its low price is the height of imprudence.

7. The account-book criteria of industrial efficiency do not accurately measure the public benefit. For example, London Transport recently asserted 'there is no economic reason' why children who occupy a seat should pay only half fare. The reason is, of course, that it has always been customary. This raises the whole question of whether an economic reason should override other reasons, and how it is measured anyway. No economic reason why a company should provide a staff canteen can be given unless we learn how to measure the better functioning of the industry as a result of better working conditions for employees. It is not an answer to say that we must learn to make this measure. One cannot try having a canteen and then abandon it if it proves to be of no economic advantage to the company. We have to make many decisions every day which are choices among alternatives, and different people may even make different choices solely for the sake of variety. In our personal lives we have to make decisions every day which are matters of personal preference and we need to have mechanisms in government by which our clear preferences can override 'economic reasons'. We are in too much of a hurry to gain economic advantage, and having gained it we merely make ourselves more hurried, and hurried by economic criteria.

Perhaps a better way would be to have a unit or ministry whose job it is to examine every decision which seems to have originated in economic reasoning and set out clearly the real consequences for the quality of our life before the decision is confirmed.

8. Politics today is very short-sighted. The social reformers of seventy years ago had objectives which were clearly not attainable in their lifetime. Some of their aims have been attained long after the pioneers died. For the world of today no similar long-term objectives are the main theme of our political parties. Government is one continual improvisation in the face of problems which are not envisaged in their immediate programmes, still less in their main policies. Governments have no energy for the formulation of new objectives except by bringing in new men who have recently done some new thinking. We are more obsessed sentimentally with the plight of the aged than with the opportunities our children, and their children, will have. The aged have no economic power, nor do our grandchildren, and so if we throw one on the scrap heap and steal capital from the other no one notices. The aged can vote, and thereby obtain some protection: the young must protest as vehemently as they can before they too become absorbed into the system as conforming adults.

A principal task for pressure groups is to set our sights further into the future, by talking about the future, by seeing how present practices commit us, by seeking ways to protect our future options.

9. Above all, the public must make politicians answerable to the people on all aspects of environmental policy. Patching up in the present is good. Every problem that presents itself must be dealt with; but that does not really require us to have policy. It merely requires that resources be allocated when the tenders have gone out and the estimates gathered in. The method of solution will usually be determined by civil servants, scientists, and engineers.

The questions the public must ask are, 'What is your party's policy to prevent an ever more frequent succession of environmental problems? How do you propose to retain room for manœuvre by ensuring a problem does not have to begin to produce dire results before its solution is given a high priority?'

We have a wage negotiation machinery, and only rarely is there a confrontation and a test of strength in a strike. In a confrontation we narrow down the room for manœuvre until one side sees that it has none left.

'What is your party's policy for preventing a confrontation between our industrial society and the environment we live in and depend on?' We need a 'negotiating machinery' which looks ahead.

These objectives may seem vague. They challenge the acquisitive society to question its own objectives. Most of the criticisms we make of its gluttony, its arrogance, its ruthlessness have been made throughout the ages by philosophers and saints, but excesses have been mollified only by the death or disgrace of the perpetrators. Usually it has been successive generations that have brought real reform. There is no new principle of brotherly love to be proclaimed, but on this occasion we may not have time to wait for the usual processes to operate.

Our children's children will find themselves overcrowded, underemployed, polluted, poor, and probably underfed, and it will avail them nothing that we, by taking no thought of our legacy to posterity, crowded them into an inadequate world, and imprisoned their lives in the abundance of people.

With what sort of pride can we proclaim that we compete more fervently than any former generation to maximise our consumption of capital resources? We shall leave our children an abundance of machinery but a scarcity of fuel to run it.

Vultures and lions

Every species has an ecological niche. Lions and condors can prey upon the more peaceful and numerous animals, who in turn live on grass.

The rag-and-bone men are the vultures of society, who live on other people's left-overs. Our trouble is that we have laws which make life too easy for the lions and unprofitable for the scavengers. The lions made the laws! It is as if they had domesticated the zebras, squabbled for grazing lands for their herds and, having over grazed, found their cubs starving.

The rag-and-bone men are going out of business at the very

10

time when man's waste products are beginning to become an important feature of the environment. We have had to institute public refuse collection because we have put the scavengers out of business by making new things and new materials too cheap, and our activities have become ecologically unsound. We have killed the scavenging business by making the working conditions unacceptable to people: our sewage is disposed of by bacteria, but most of our rubbish accumulates, and the only remedy is to produce less of it.

We need a greater reverence for Nature, a clearer picture of man's cosmological insignificance. After evolving for hundreds of thousands of years to the present level of scientific intelligence, the short-term prospect we have created makes it almost irrelevant to think of a real future. Heaven is irrelevant. The best investment bargains have a seven-year term, and almost none of us is sacrificing present profit and giving his creative effort to the long-term future. The young who take out a mortgage are not thinking of their old age but of the best subsidised rent bargain.

Yet this has not been the wish of scientists, nor of the common man. What can we expect when most newspapers devote more pages to methods of getting richer without working for it, or at the very best by working only for one's self, than to the encouragement of any other activity? It is all based on the assumption that by some mysterious means the manipulation of money which receives the greatest reward serves humanity best. Since the reward has come from mineral wealth stolen from posterity we need laws which do not encourage this maximisation of theft.

Science and the obvious

The Secretary of State for the Environment said the first priority of his department was to improve the environment of those who had a bad one at present.

The first task was to develop as complete a monitoring system for pollution as possible. Only when that was achieved and the results published could the public assess progress in reducing all forms of pollution.

It was intended to publish regularly updated river

surveys so the public could trace the improvement taking place in the quality of the rivers.

On air pollution, he said a complete survey had been made by the Warren Spring Laboratory. Later this year the findings of the analysis would be fully published.

The Times, 4 March 1972

Scarcely concealed here is the insidious idea that we do not know enough about pollution to decide what to do about it until we have made extensive measurements of it. The idea is dangerous because it is false. It is the product of a system in which scientists at work are held apart from the public at large by a lying mystique that the ordinary educated man cannot involve himself in issues. Merely to express an issue in scientifically quantified terms does not mean that it can only be understood that way; indeed issues are often obscured that way.

In arguing about these issues we shall surely find the scientist who will argue that science is making quantitative measurements, expressing things in terms of numbers, and so on. What he is trying to justify is his assumption that all things are best understood that way, and must be argued about in such scientific terms. And more: if they are not understood that way, they are not understood at all.

Such technocratic arrogance must be exposed. It is irresponsible for government scientists to pontificate on public issues in their own narrow jargon. They do it in part because it protects them from public criticism, in part because they are not compelled to think in terms of direct communication with the man in the street. But pollution policy must be intelligible, first of all to elected representatives both at the parliamentary and local government level, and these representatives are not specialists.

In the case of air pollution there do exist, in Britain, many ways in which we are protected from technocracy. For example, local councillors and public health inspectors operate at the level of the man in the street and at the same time have to be briefed by the scientific civil service. Such organisations as the National Society for Clean Air provide a meeting point for all concerned to argue the issues without the inhibitions of professional or political pressures. There is also a 'Conference of Co-operating Bodies' at which members of the local authorities who participate in the National Survey of Air Pollution meet together with scientists of the Warren Spring Laboratory (the government establishment in which the scientific civil servants meet). There is a 'Clean Air Council' presided over by a minister in the government, who is a politician and probably not a scientist, at which scientists, industrialists, and local councillors meet to discuss policy issues.

The comment appropriate to the minister's statement to the House of Commons, quoted above, is that the minister himself seems to think in terms of a technocracy and encourages this separation of the science from what ordinary people experience. Perhaps it is designed to protect him from lay discussion in the House. The opposition front bench is not perturbed because it saves them the effort of getting themselves properly briefed on technology.

The success of the 1956 Clean Air Act is really measured by what we can see with our eyes and smell with our noses. It is blatantly obvious to anyone who has tried to observe, who remembers what London was like before, or who can be bothered to sniff and look in towns where no air pollution control has been carried out. It needs only an angler to inform us of the quality of water, and it needs only a public health inspector to find with his eyes most of the main sources of air and water pollution.

The technological challenge is not to monitor but to eliminate pollution. There need be no detective problem; we can know perfectly well what and where it is before it is emitted. Every industrial emission could be public knowledge by law, whether it be to air, to water, or to the land. If an industrialist wants to keep his wastes secret he should keep them on the premises. We must not let ourselves be misled by the obvious need to control many special cases by special monitoring and expensive equipment, into the belief that most pollution needs to be controlled that way. Radioactive pollution, asbestos dust, beryllium, fluorine, and quite a long list of other substances emitted by industry should not be tested by sniffing and peeping; they need very great care. But the need for that special treatment need not slow down our general, and much more important, anti-pollution work. Ministers, and those who debate the matter in the legislatures, must get themselves as well informed on the technical aspects of the subject as they are on financial and legal matters when they debate them. This can only happen if scientists, too, take on new attitudes and deem it to be their responsibility to make any important issues intelligible to the public.

The purpose of education is not so much to train technicians to run our technological civilisation as to equip us all to understand what is going on.

Is there a simple message?

Here, at the end, reviewers will hope to find a message to save them from the need for deep thought on the big issues discussed throughout the book, and to which this author can see a satisfactory solution coming only at the end of long discussion and experimentation by enterprising communities. What we need is a theme around which society as a whole can educate itself and adjust its attitudes so that it will develop collective instincts and aspirations appropriate to the full world.

The whole book is, one hopes, such a message, but let us simplify it as a starting point for the long discussions.

Henry Ford regarded it as wasteful for one part of a car to last longer than any fundamental part. If all its parts did not fail together some had been made too expensively. The whole vehicle should suddenly collapse together. His ideal was to make good cars as cheap as they could be. Much of the design of modern cars is based on ease of production and ease of maintenance, but never upon ease of salvage. If the purchaser were compelled when he bought a car to pay, not for its ultimate disposal as rubbish to be got out of the way, but for its complete salvage and re-cyclement, then what? To replace Ford's ideal car, we now need one which, at the push of a button, will disintegrate into separate piles of glass, steel, rubber, brass, aluminium, leather, etc., ready for further use as raw materials. The paint and other aspects of the design should be such as to make re-use easy. The pile of material which cannot be re-used should be highly taxed.

A criterion of this kind should be applied to every industrial product.

In ways like this we could build into our activities serious consideration for the future. Perhaps the most ludicrous aspect of our price structure is that minerals are priced according to the cost of obtaining them. This seems natural to the average orthodox economist because it is the way we have always treated farm produce. The farmer, however, used to plan that his farm should be the same at the end of the year as it was at the beginning so that production of food could go on indefinitely. Unfortunately today even farmers' practices are becoming ecologically unsound and now depend absolutely on mineral fuel for harvesting and cultivation and for the production of their artificial fertilisers.

Brass tax

Economists think too much about money. To get revenue they look around for money, and propose that a percentage of it should be surrendered as tax. They tax the 'brass'.

But when you get down to fundamentals, it is obvious that in taxation policy the mere raising of revenue will become subsidiary to the much more important objective of planning human life and ensuring freedom in the full world humanity will soon find itself in.

We have scarcely learned to appreciate the climax of the Industrial Revolution although we are already at the beginning of a new era in human evolution. At such a moment we must look forward to a new, cleaner, less extravagant technology; that alone can keep us free when we circumscribe our lives by our own multitudes. The message we need is not to be found in history.

The nuclear age

If we look forward into the future and see what pressures we shall experience, not from economic causes but purely physical ones connected with the limitations of the world, we should now seek to anticipate them so that they will come into operation more gradually and we will have room for manœuvre while we are learning to meet them. Appropriate action would be to place our tax burden on fuel, instead of at convenient points in the accounting business. The consequences of a fuel tax will be great, many and unpleasant to accept in the short term; the tax would not be much use otherwise.

Every brick, every piece of metal, plastic, glass, paint, every ounce of fertiliser, every cooked meal is produced by the expenditure of fuel. Our trains, drains, heat, light, and every piece of manufacturing machinery can only be used by expenditure of fuel. Fuel has replaced man-power as the source of wealth many times over in all industrial societies. As long as fuel is plentiful that seems fine. But now its exhaustion is threatened our main tax burden should be placed upon it because it is our main source of wealth; only that way will it be used more efficiently and less extravagantly; only that way will conservation become a main political theme.

Whenever the imminent shortage of fossil fuel is discussed there are always optimists who say 'nuclear energy will solve all fuel problems'. It is quite probable that they are right in the long run, but there is a simple-minded optimism abroad which makes people believe that with enough effort put into research and development we can achieve anything that is theoretically possible. In fact, we are a very long way from cheap nuclear energy: at present no useful energy is produced from nuclear fusion at all, and although about one-sixth of Britain's electricity is generated from fission energy, Britain is far ahead of the rest of the world in this and produces half the world's total. But Britain's total energy consumption is only one-thirty-fifth of the world's.

We could perhaps subsidise the development of nuclear electricity generation artificially in order to promote more rapid progress, but the difficulty about subsidies in our accounting system is that although we know in our hearts that we are entitled to spend our money how we please, we still believe that the account books should look right. Since subsidies look inefficient, other methods may be needed if a long-term programme is to be successful. It would be most unsatisfactory if those engaged in the developments had to keep persuading the politicians every five years to continue the subsidy for another five when the developments themselves may require a thirty-year investment programme.

If the price of fossil fuel were raised by taxation and that became our habitual source of revenue for the next century, as income tax has been for the last, nuclear generation of electricity could now show a handsome profit which could be used for further expansion.

But let us not be simple minded about it: there are also some serious disposal problems for radioactive wastes. These are not large at present because so small a fraction of our total energy is of nuclear origin. Atomic fission energy is, furthermore, another sort of mineral fuel, and the cost and difficulty of mining uranium will increase as the easily accessible deposits are used up. Nuclear fusion energy has not been tamed (a megaton bomb is not tame) and what will be the radioactive by-products, if it is tamed, still remains to be seen. We simply cannot afford, for our children's sake, to be complacent about nuclear energy: the dawn of the nuclear age is not yet even the palest grey.

Such is the situation that every alternative source of energy is being actively considered – sunshine, temperature gradients in the sea, hot springs, estuarial barrages and, of course, the more economical use of all energy. But many developments of this kind are frustrated by their apparent high cost. The Severn barrage scheme would have been a very good investment when it was first proposed in the 1920s, but now it is far too costly as an energy source because oil is so cheap, and it looks like a very bad investment.

We are only slowly emerging from the Industrial Revolution in which muck and money went together. Fuel efficiency and fuel economy are the best allies of clean air.

Index